THEORY
OF
EVERYTHING

THEORY
OF
EVERYTHING

Efren Basa Aduana Jr.

THEORY OF EVERYTHING

Edited and published by Efren Basa Aduana Jr.
Printed by CreateSpace

ISBN-13: 978-0615888187
ISBN-10:0615888186

For my mother, Rachel

Master of Matter

by Efren Basa Aduana Jr.

Count the number of subatomic particles.
Three: proton, neutron, and electron.
Quick! Now count the number of fundamental particles.
Five? Six? Seven? Eight? As many as the stars?
Most will agree it's bewildering,
None can come up with a good understanding.
Even the many great luminaries of physics,
Cannot divine nature's basics.
Where did the mass of the particles came from?
That to them is a great conundrum.
But time has come when the mysteries unfold,
When the new ideas replaces the erroneous old.
Let us then herald a brand new future.
Set our minds on a new venture:
To master matter,
To know ourselves better,
To make the world better,
And to make us kinder towards each other.

Contents at a Glance

Table of Contents

PART 1: THE ATOM

PART 2: THE THEORIES

Table of Contents

PART 3: REAL NEW PHYSICS

Preface

The title of this book is "Theory of Everything." It was not titled as such like other physics books to discuss how hopefully the Theory of Everything could be attained or what theories are vying for the Theory of Everything or just to make the reader think that there is in this book a "theory" of the Theory of Everything—this book actually contained the Theory of Everything. (Refer to Appendix D for my original Preface.)

Lest it will be drowned in the popular scientific endeavors of this day (such as the search for and discovery of the Higgs boson, dubbed as the "God particle," which is said to be the particle that gives mass to other particles), this book forwards the following theories: New Model of an Atom, Theory of the Structure of the Atom of the Elements, Atom Law (to replace the Pauli Exclusion Principle), Quark Theory (the structure of quarks inside the proton and neutron), Charge Theory, Mass Theory (theory on the source of mass of the particles), revisions on the Standard Model, and the Theory of Everything. I wrote this book not to disprove the existence of the Higgs boson—rather that the understanding of the source of mass of the particles was just a consequence of my discovery of the basic structure of the atom forwarded in the New Model of an Atom. What I am bringing about to science is what everyone had been looking and waiting for. It will be a shame if people started to get hung up on the issue of the question of the existence of the Higgs boson as there is more than that that my theories and ideas in this book can offer. This book is overthrowing established theories and even the Nobel Prizes that were won. In short, this book will bring forth no less than a paradigm shift.

As I am gaining knowledge in physics, I had the opportunity to know which of some of the current theories and ideas are right and wrong. As such, I am dismayed by what is happening in science right now, particularly in the field of particle physics and astrophysics. It is like the story of the "Emperor's New Clothes"—the empirical science is naked. The scientists truly believed their theories and ideas are right. But to me, science had become a farce—to be frank about this—as I know that some books, articles, and pronouncements in science were wrong. But for what is happening in science right know, I don't blame the scientists at all; it is just that it was not the time yet for the true knowledge to come out. But the time for correction is *now*. As I have something to offer to science, I believed it is my moral and intellectual obligation to correct this current situation to point science to a new direction. There is just too much brain, time, and money being wasted throwing wild theories and ideas and chasing after them. It is my wish and intention that science should be more practical and of service to the immediate needs of humanity. In this climate of change, it is a moral imperative for those people who are intellectually endowed to do something to protect our world.

Physicists need great and purposeful endeavors. There are exhilarating knowledge to be gained and discoveries to be achieved. Let us move forward towards the new age of physics.

Efren Basa Aduana Jr.
Arlington Heights, Illinois
August 2013

Introduction

What is the Theory of Everything?

Simply speaking, the Theory of Everything is a theory about how the world was created, its basic building blocks, and how it operates. It implies an explanation of all what is contained in our universe in a single simple explanation. The prevailing theory right now is the Standard Model, which is about the fundamental particles (family of electron called leptons, and quarks inside the proton and neutron) and the fundamental forces of nature (electromagnetism, gravity, and the strong and weak forces of atom).

The search for a single, simpler explanation of all the things in the universe also led to the search to *unify* all the different fundamental forces. The first idea of the unification of the forces came from the unification of electricity and magnetism by James Clerk Maxwell in his Electromagnetic Theory. This was followed by Albert Einstein who, taking upon the idea from Maxwell, tried to unify electromagnetism with gravity through his General Theory of Relativity but failed. The next recognized unification was that of electromagnetism and weak force, called Electroweak Theory, by Steven Weinberg and Abdus Salam.

As of now, physicists are still in the quest of unifying all the forces, whether in separate parts or as a whole. In all of these, gravity provides a stumbling block as it eludes any idea on how to unify it with all the other forces. As such, the Standard Model is also called as the Theory of *Almost* Everything in the sense that it cannot explain gravity. Currently, the Standard Model is also being argued that it cannot explain the idea of dark matter and dark energy. As such, the physicists are already talking about what is beyond the Standard Model.

There is another theory that is trying to be the contender for the Theory of Everything—the String Theory. The idea of the String Theory is that all the different fundamental particles of the Standard Model are really just a manifestation of a string. That is, where the Standard Model views the particle (such as the electron, quarks, and bosons) as zero-dimensional or point-like object, the String Theory posits that the particle is actually one-dimensional in the shape of an oscillating string. The String Theory is said to "allow for the consistent combination of the Quantum Field Theory and General Relativity Theory that agrees with the general insight of the current theories of Quantum Gravity."

What can be said about the physicists' understanding of the Theory of Everything is that it can only be attained by pure mathematics—with the idea that it can be printed in a t-shirt much like Einstein's famous equation, $E=mc^2$.

Organization and Highlight of the Book

This book is organized into three parts in eight chapters.

- Part 1 is composed of two chapters, which are a background on the model of an atom and the Standard Model, discussing their history and how their knowledge was achieved.
- Part 2 is composed of five chapters, which are about the true understanding of charge, the new model of an atom, the structure of the quark inside the proton and neutron, and the application of the new model of an atom to the structure of the elements in the Periodic Table of the Elements, and on chemistry.

- Part 3 is a one chapter on the right understanding of the Standard Model, reorganization of the fundamental forces, and the Theory of Everything.

At the end of Chapter 1 and Chapter 2 is a *Note* of what the author would like to point out to the readers regarding the subject of the chapter and what the readers need to remember in the succeeding chapters. At the end of Chapter 3 to Chapter 7 is *Summary* of each chapter to recap of the discussions of the chapter.

Author's Approach to Writing This Book

I meant for this book not just to challenge the current knowledge in physics but also to provide theories that could correct or replace those existing theories. Aside from the organization of the table of contents and the presentation of my ideas, the reader could, with a keen observation, somehow deduce how I had reached the solutions to the problems of physics in this book.

Comparison with Other Physics Books

Academic books in physics discuss the background and knowledge of physics with the idea that students should be exposed as much as possible to all the knowledge of physics. Some books in physics, especially by famous physicists, discuss the current developments in physics with the idea of inspiring interest in their field of study or to advance their theory or idea they support. There are some physics books that challenges the dominant ideas and theories of physics (such as the Big Bang Theory and the String Theory) that garnered opposition as they don't have enough strong argument to defeat these dominant theories or they don't have a new theory to replace the current ones. Where the latter fails, I feel that this book will succeed.

Theories and Ideas

The theories and ideas I had forwarded in this book could be original or supported by existing but sometimes unrecognized theory or idea. But most of all, the theories agree and support each other. This is a complete departure from the current approach of science that in order to solve a complex problem, the problem should be solved in parts—which oftentimes produces theories in disagreement with each other (such as the case with Quantum Mechanics and General Theory of Relativity).

Method of Solving a Problem

A search for the scientific method of solving a problem shows that each problem requires different procedures. A typical scientific method constitutes the following steps:

1. Formulation of question
2. Hypothesis
3. Prediction
4. Testing/Experimentation
5. Analysis/Final statement

The method above is usually for those who are already working in a field of study. What I had learned from inventing in coming up with an idea or solution are the following: you have a "sudden" inspiration; you can invent if you are working in a certain field of study or work; and that you think of a field, find a problem, and then invent. The same is true with the latter on what I had done in writing this book. The following are the steps I had done in solving the problems in this book:

1. Learn of the existing problem
2. Research the problem
3. Think the solution and formulate a hypothesis or theory
4. Test the hypothesis or theory

5. The theories should agree with each other

In step 5, I had to rely on my understanding of which existing theories are right or wrong. In all of this, it helps to be open to the possibility that the existing theories could be wrong and to be healthy skeptical with everything I read. Obviously, I rely on my foundational scientific knowledge and intuition to work side by side to guide me through the probable solution to the problems that I will encounter. Solution to the problem can be achieved through brute rationalization of the problem, which makes the brain work it out even days, weeks, or months later, even when I am not actively thinking about it. There is also the uncanny and lucky encountering of some related articles, inspiration, or intuition that works beyond my belief.

Who This Book is For

This book is for those ordinary people who are interested in science and the current scientific developments; for those who are in the academe: students, teachers, and those taking graduate studies; and those who are scientific researchers, and theoretical and experimental particle physicists. Underlying every theory is still a basic and easy to understand idea that does not need the complex mathematics to explain it.

Ordinary people, the students, and the expert in field of physics will benefit in this book as it is a return to the basic foundation of this field of physics that will be a springboard for the future.

On Style and Capitalization

This book was loosely based on Chicago Manual of Style (CMS): caption for figures and tables, the footnote, and bibliography follows the rules of CMS. However, inline reference to figures and tables, which are in lowercase in CMS (ex: figure 2.1), for my reason of being lost in the sentence, are capitalized (as I also found rather *pleasing* in some books I had read). Also, for personal reason of making the term stand up, the following examples are the changes in the accepted capitalization in this book:

Standard Rule	In This Book
physics	physics
classical physics	Classical Physics
quantum mechanics	Quantum Mechanics
quantum electrodynamics	Quantum Electrodynamics
Big Bang theory	Big Bang Theory
string theory	String Theory
photoelectric effect	Photoelectric Effect
Pauli exclusion principle	Pauli Exclusion Principle
particle theory of light	Particle Theory of Light

Part 1
The Atom

Chapter 1
The Atom Model

The Ancient Philosophers

In ancient Greece there were two different schools of thought regarding the nature of things: one is that matter is continuous and that it is composed of four fundamental elements: fire, air, water, and earth; and the other is that the world is made of small indivisible particles called *atom*. Atom came from the Greek word "atomos," which means "uncuttable" or "indivisible."

Empedocles (490–430 BC) was the earliest known supporter of the idea of the four elements. Plato (429–347 BC) espoused the same idea but goes further by adding that the four elements are in geometric forms. Aristotle (384-322 BC), a student of Plato, supported the idea of the four elements but also added a fifth element called *quintessence*, which he thought of as a divine substance that makes up the heavenly bodies and also ties all other elements together. Aristotle's writings influenced much the succeeding civilizations (flourishing to the time of the Renaissance and the time of Isaac Newton).

The idea of the early atomic theory was credited to Leucippus (born around the half of the 5th century BC) and his student Democritus (460–370 BC). Both of their works were difficult to separate from each other, although Democritus came down in history as the more famous. Leucippus and Democritus believed that the world is made up of atoms.

The following is the timeline of the Greek philosophers:

- Empedocles (490–430 BC): Four elements
- Leucippus (before 460 BC): Atom
- Democritus (460–370 BC): Atom
- Plato (429–347 BC): Four elements
- Aristotle (384–322 BC): Four elements and quintessence

The timeline of the ancient Greek philosophers shows that there was the handing down of ideas or beliefs from the teacher to the student and that to some extent there could be interactions between the men living at the same time. For example, Democritus who was largely ignored in ancient Athens was known to Aristotle, who was said to disliked him so much that he wished all of his books burned.[1]

The Chemists

With the ancient Greek's interactions with Mesopotamia and ancient Egypt, metalworking took a new importance. The discovery and use of the early metals such as gold and copper led the civilization of the ancient Egyptians and the Mesopotamians into greater developments. Ancient Egyptian metalworkers knew how to make imitation gold by mixing copper with other metals. The discovery of ever harder metals such as the mixing of the copper and tin to make bronze and the subsequent smelting of iron led to superior weapons to both conquer and expand their lands.

Macedonia under Philip II defeated the Greeks to consolidate his force with the plan to attack the Persian Empire. With Philip's assassination by his bodyguard, his son Alexander III came into the throne. Alexander III (Alexander the Great) subsequently conquered Egypt, Persia, and many other kingdoms and reaching as far as India. With Alexander's death in 323 BC, ancient Egypt came under the dynasty of Ptolemy, one of Alexander's general. The dynasty of Ptolemy ended with the victory of Octavius (grand-nephew of Julius Caesar, who later was known as Augustus) over the Queen Cleopatra (Ptolemy) and her lover Antony. With the death of Cleopatra in 30 BC, ancient Egypt became a Roman province.

The fall of the Roman Empire by the end of the 5th century led to the period called the Middle Ages when the Western civilization came into stupor as it seems the light of knowledge had dimmed, hence it was also called the Dark Ages.

With the birth of Islam in 622, the Muslims rapidly expanded their conquests. In 640, the Arabs conquered Egypt and became the inheritor of the Greek, Roman, Mesopotamian, and Egyptian knowledge. The word "chemia" was the root word of chemistry. The Arabs took up the study of "chemia," which came to be called "al chemia," being that "al" in Arabic means "the." "Al chemia" is also the origin of the word "alchemy."

The Muslims reached as far as Spain and Portugal before their advance was stopped in 732 in France. Spain was under the Muslim rule for the next 700 years. While Europe was in a state of stagnation, Islam saw the flowering of its intellectual pursuits. During this time, only the Christian monasteries were the center of knowledge. Monks took the task of translating the Arabic works into Latin.

In the early 1400s, a movement called Renaissance (meaning "rebirth") started in Italy, which then spread all over Europe. Their hunger for knowledge led to the study of the earlier works of the Greeks and Romans, as well as the works of the Islamic scholars. Scholars became interested again to read the surviving works of the ancient philosophers.

The Alchemists

The early alchemists were preoccupied with the mixing of different substances, particularly the idea of transmuting base metals such as lead into gold. (Isaac Newton was also known to have secretly studied alchemy, having himself a copy of the *Emerald Tablet*, which purportedly contained the description of how to turn lead into gold.)

Subsequently, the practice of alchemy fell into disrepute as some alchemists' leaned towards mysticism and others into fakery. Alchemy then was relegated into quackery and witchcraft with its critics buoyed by the idea of a person secretly boiling mixtures in a cauldron in the dead of the night. Alchemy had become synonymous to pseudoscience.

Paracelsus (1493-1541), a German-Swiss doctor also practiced alchemy. He was not an ordinary alchemist in a way that he was not preoccupied with finding the method of turning lead to gold, but rather he believed that alchemists should search for medicines to cure diseases. He wanted to end the ancient belief of alchemy through the practice of methodical experiment.

The foundation of modern chemistry was laid by George Bauer (1494-1555), a German scholar and scientist in the 16th century, with the development of systematic metallurgy, the extraction of metals from ores.[2] He is known as the "Father of Mineralogy."

The Modern Chemists

The desire to make accurate measurements could be traced back to the ancient study of astrology and astronomy. Both were practically indistinguishable in the early times. During the Renaissance period, Galileo (1564-1642), an Italian physicist made accurate measurements in physics and astronomy that would establish the practice of accurate measurement from the experiment.

Robert Boyle (1627-1691), an English scientist greatly admired Galileo and his precise method so that he would apply it to the field of chemistry. Boyle observed that if he trapped some air and put pressure into it, he found that if he measure the pressure and double it, the air was forced into half its original volume. If he tripled the pressure, the volume of air measured would be a third of the original. That is, the pressure of a given mass of (an ideal) gas is inversely proportional to its volume at a constant temperature. This discovery is still called *Boyle's Law*. In 1661, the subsequent publication of his book, *The Skeptical Chymist*, challenged the early Greek's idea of the four elements of nature, while at same time argued for the virtues of the atomic theory. The idea through experiment took hold that if a substance could be divided into simpler part, it was not an element; and likewise, if it could not be divided, it is an element.

In the 1700s, the idea of element had been tested when George E. Stahl (1659-1734), a German chemist advanced the idea of an invisible substance called *phlogiston*. He forwarded the idea that anything that contained phlogiston would burn. Wood and coal, which were full of phlogiston, would burn and as they burn the phlogiston would pass out into the air. The remaining ashes left from the burning of the wood or coal lacked the phlogiston and could not burn anymore. He also thought that metals were full of phlogiston. When metal rust, it would lose phlogiston to the air so that the rust produced would lack the phlogiston.

Antoine Lavoisier (1743-1794), a French chemist finally was able to challenge the idea of phlogiston. The practice of accurate measurements brought about by Galileo and Boyle led him to produce exacting measurements from his experiments. With the idea of the phlogiston, when a wood burned it should lose the weight since it loses the phlogiston, while on the other hand when the metal rust he observed that it had gained weight. There was clearly something amiss with this observation. He experimented by heating a metal in a closed container. He observed that although the metal rusted, the container and the metal did not change in weight. If the metal gained something when burned, then something in the container must have lost some weight to balance the gain. The natural source of this added weight that can be easily observed is the air. That is, the metal is not losing the phlogiston when it rusted but rather it is gaining something from the air. He came into conclusion that phlogiston could not be involved and that instead, air is made up of elements. He surmised that the air must be made of two gases. He called the gas that burned as *oxygen* and the other gas that does not burn as *nitrogen*. Lavoisier stipulated what became the *Law of the Conservation of Matter*, which states that matter is neither created nor destroyed but only changes its form. He, along with the others made up the naming of the chemicals that is still in use today. In 1789, Lavoisier published his book *Elementary Theory on Chemistry*. (His expensive apparatus and his exacting experiment were carried through with his excessive practice of tax collection and was one of the reasons why he was subsequently beheaded during the French Revolution.) For his work, Antoine Lavoisier is called the "Father of Chemistry."

Following the works of Lavoisier, subsequent scientists performed careful experiments to study chemical reactions and the composition of various chemical compounds. Joseph Proust (1754-1826), a French chemist showed that a given compound always contained exactly the same proportion of elements by mass. This principle became known as the *Law of Definite Proportion*.

John Dalton (1766-1844), an English schoolteacher inspired by the work of French chemists revived the concept of the atom based on the facts and experiments. He reasoned that if the elements were composed of minute particles called atoms, a given compound should always contain the same combination of these atoms, which explains why the same relative masses of elements were always found in a given compound. In 1808, he published his work *A New System of Chemical Philosophy*, in which he presented his new atomic theory. The theory can be summarized as follows:

- Each element is made up of minute particles called atoms.
- All atoms of a given elements have identical properties (mass and size), while atoms of different elements have different properties.
- Compounds are formed by the combination of the atoms of the different elements.
- Atoms combine to form compounds in simple numerical ratios.
- Atoms of different elements may combine in more than one ratio to form the different compounds.

Slowly but surely, chemistry was ready for a new development—a somewhat natural occurrence in the field of science when the dam of ideas is ready to burst.

The Periodic Table of Elements

As more elements, in which some were gas and some were metal, were discovered during the 1700s, the problem arose of why the elements were different from each other and similar in number of ways, yet they could not be changed into each other.

By the 1820s, Jöns Jakob Berzelius (1779-1848), a Swedish chemist analyzed many substances carefully making measurements of how much of each element are in each substance. His experiments yielded quite accurate atomic weights. Berzelius also suggested that each chemical element should be given a symbol starting with its first and sometimes second letter. This was an improvement from Dalton's method of using geometric symbols for the elements.

More and more elements were later discovered having properties of metal and gas. Chemists tried to arrange the elements according to their atomic weights to perhaps make sense of their list. In 1869, the breakthrough came from Dmitri I. Mendeleev (1834-1907), a Russian chemist who arranged the elements into rows and columns according to their atomic weights, where the arrangement showed the elements grouped with similar properties. Mendeleev's arrangement was called the *Periodic Table*. To make the elements fit properly, he had the foresight to leave some places empty; predicting that these empty places contained some elements that are still to be discovered. To show the correctness of his arrangement, he picked three of such elements and figured out what their properties would be. His predictions matched the elements Germanium (Ge), Gallium (Ga), and Strontium (Sr) when they were discovered in the 1870s and 1880s.

In 1913, Henry G.J. Moseley (1887-1915), an English physicist arranged the Periodic Table in a new way. He noticed that if the elements were arranged according to their atomic number (the number of protons, which is also equal to the number of electrons), the elements in vertical column showed similar properties as well as some of those in the rows. Hydrogen, which is the first element stood apart since its chemical properties are not like those of the other elements. With Moseley's contribution, the Periodic Table of the Elements came to its modern form.

The Physicists

The basic foundation of chemistry and physics regarding the atom is very much intertwined. And so, as the chemists were the ones who had ushered the revival of the atomic theory, it is just but natural that the physicists were the next one who would take a stab at the atom. It's just a matter of time before the physicists would catch up.

Famous atomists René Descartes (1596–1650) and Isaac Newton (1643-1727), who were both physicist and mathematician, harbored the ancient Greek's idea of atom in the idea of the particles called "corpuscles," which they accorded to the nature of light. For a time, the Particle Theory of Light had been very influential.

Also, during the time of Newton a new theory of light called Wave Theory of Light was taking root. Robert Hooke (1635-1703), who Newton had a great misunderstanding, supported the Wave Theory of Light. They had disagreement for a long time so much that Newton did not publish his work until Hooke's death in 1703.

It was with the two-slit experiment of Thomas Young (1773-1829), an English scientist that started the erosion of Newton's Particle Theory of Light. The subsequent works of Augustin-Jean Fresnel (1788-1827) completed the ascendance of the Wave Theory of Light over the Particle Theory of Light.

Thus the nature of light succumbed to the idea that light is a wave. It would take some time for the Particle Theory of Light again to resurrect with the work of Einstein in his Photoelectric Theory and subsequently the revival of the atomic nature of matter.

The Discovery of Electron (1897): J.J. Thomson

Cathode ray tube is an evacuated glass tube with two electrodes at both ends where a voltage is applied. Cathode rays are streams of electrons coming from the cathode or negative end towards the anode or positive end of the cathode ray tube (Figure 1.1).

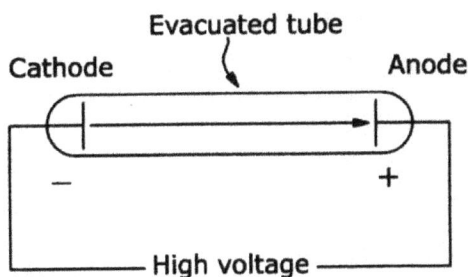

Figure 1.1. A cathode ray tube.

* * *

Following the invention of vacuum pump in 1960 by Otto von Guericke (1602-1686), a German scientist, inventor and politician, physicists started to experiment with passing high voltage electricity through the rarefied air.

In 1838, Michael Faraday (1791-1867), an English scientist noticed a strange light arc emanating from the cathode moving towards the anode when he passed a current through a glass tube filled with rarefied air. In 1857, Heinrich Geissler (1814-1879), a German physicist and glassblower evacuated more air with an improved pump and found that instead of arc, the tube is filled with glow.

By the 1870s, William Crookes (1832-1919), a British chemist and physicist, and others were able to evacuate tubes at lower pressure. These tubes were called *Crookes tube*. Michael Faraday was again the first to notice a dark space just in front of the cathode where there was no glow. This was called "cathode dark space," "Faraday dark space," or "Crookes dark space." Crookes observed that as he pumped more air out of the tubes, the dark space spreads from the cathode towards the anode until the tube was totally dark except at the anode end where the glass begins to glow.

In 1869, Johann Hittorf (1824-1914), a German physicist was the first to realized that something must be travelling in a straight line from the cathode to the anode as the glow cast a shadow when it is interrupted. In 1876, Eugen Goldstein (1850-1930), a German physicist called this the *cathode rays*. (In 1886, Goldstein also discovered the *anode rays* when he perforated the cathode and observed that the rays coming from the anode passes through the holes or channels in the cathode. He called the anode rays earlier as *canal rays*.)

Earlier in 1858, Julius Plücker (1801-1868), a German mathematician and physicist investigated the light emitted from the Crookes tube and the influence of the magnetic field on the glow.

It took a while for everyone to know the significance of the cathode rays.

In 1897, John Joseph "J.J." Thomson (1856-1940), an English physicist demonstrated that cathode rays are deflected by an electric field (Figure 1.2) as well as by a magnetic field. Since the cathode rays were produced on the negative electrode (cathode) and were repelled by the negative pole of an applied electric field, he postulated that the ray were a stream of negative charge particles. The negative charge particles were later called *electrons*.

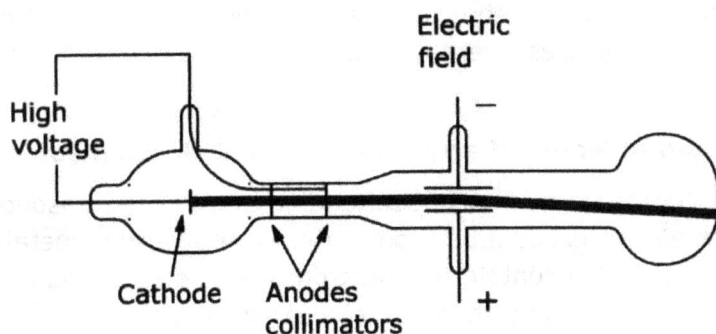

Figure 1.2. The cathode ray tube in experiment with an applied electric field.

* * *

Earlier in 1874, G. Johnstone Stoney (1826-1911), an Irish physicist proposed a concept of a unit of charge that is found in the experiments when an electric current passes through chemicals. He initially called it "electrine."[3] By 1891, he changed it to "electron."[4] It was in this sense that Joseph Larmor (1857-1942), who was J.J. Thomson's classmate, devised a theory of "electron" that described it as a structure in the ether. Larmor's theory did not describe the "electron" as part of the atom.

Thomson in 1897 originally presented three hypotheses on his cathode ray experiments. One is that the cathode rays are charged particles, which he called then as "corpuscles," reminiscent of Newton's corpuscles. Two is that these corpuscles are constituents of the atom. Three is that the corpuscles are the only constituents of the atom.[5]

When Thomson's cathode ray experiment was known in 1897, Stoney's nephew, the physicist George Francis Fitzgerald (1851-1901) suggested that Thomson's "corpuscles" were really "free electrons" alluding rather to Larmor's kind of theory of "electron" in disagreement with Thomson's hypotheses.[6]

(According to another article, Fitzgerald realized that Thomson's "corpuscles" were actually Stoney's "electrons."[7])

Plum Pudding Model of an Atom (1904): J. J. Thomson

Thomson having gained an understanding of an atom reasoned that since electrons could be produced from the various metals, all atoms must also contain the electron. Also, he assumed that since an atom was known to be electrically neutral, an atom must also contain some positive charge. Thomson postulated that a *spherical* atom is consisted of a diffuse cloud of positive charge with the negative charge electrons embedded randomly. His model became known as the Plum Pudding Model of an Atom (Figure 1.3).

Saturn Model of an Atom (1904): Hantaro Nagaoka

In 1904, Hantaro Nagaoka (1865-1950), a Japanese physicist forwarded a model of an atom where a very massive positive charge nucleus concentrated into a central core and the electrons revolves around the nucleus bound by electrostatic force. That is, the nucleus is the one pulling the electrons into circular orbit. The analogy is that of the rings in circular orbit revolving around Saturn, which are bound by the gravitational force (Figure 1.4).[8]

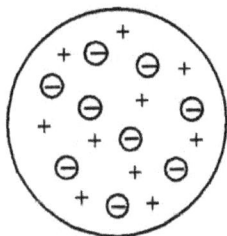

Figure 1.3. Thomson's Plum Pudding Model of an Atom. The atom is made up of a positive charge sphere and negative charge particles.

Figure 1.4. Nagaoka's Saturn Model of an Atom. The center of the atom is made of large positive charge nucleus and the electrons orbiting around the center much like the rings of Saturn.

The problem with Nagaoka's atomic model was that he could not explain how such system could be stable. Mechanically, such planetary model in which the electrons revolve around the nucleus is stable. Argument against it was that according to Maxwell's Theory, an accelerating electron would emit radiation. This loss of energy means that the electron would spiral into the nucleus within 10^{-8} second. This obviously does not happen.[9] Nagaoka himself abandoned his theory in 1908.[10]

Brownian Motion and the Atom (1905): Albert Einstein

In spite of the chemists' assertion of the existence of the atom, physicists were still skeptical in accepting it. The rise of an intellectual giant started with proving the existence of the atom.

In 1905, Albert Einstein (1879-1955), a German-born theoretical physicist who later immigrated to the United States published, his paper to explain the Brownian Motion.[11] In it he described the Brownian Motion as due to the molecules of water crashing at random into the suspended particles. His paper had an important role in securing the acceptance of the atomic theory by the physicists.[12] Einstein explained that through molecular kinetic theory of heat, the thermal molecular motions of molecules could be observed easily under the microscope.

Rutherford's Model of an Atom (1911): Ernest Rutherford

Ernest Rutherford (1871-1937), a New Zealander studied subatomic particles and earned his doctorate working for J.J. Thomson. After his graduation, he went to McGill University in Canada and began his work in the field of radioactivity. A year later, he went back to New Zealand, got married and returned to Manchester University in England continuing his study of radioactivity. In 1899, he discovered and coined the term "alpha rays" and "beta rays" for the two types of radiation. Alpha particles came from the atom of helium (two protons, two neutrons, and two electrons) with stripped electrons, while beta particles are high-speed electrons.

In 1900, Paul Villard (1860-1934), a French chemist and physicist found a third kind of radioactivity that Rutherford had missed because it passes unnoticed on his apparatus. Villard did not suggest a specific name for the particle. It was Rutherford in 1903 who proposed to call Villard's rays as "gamma rays" as they were more penetrating than the alpha rays and beta rays he had previously discovered.[13]

In 1909, Rutherford asked his two assistants, Hans Geiger (1882-1945) and Ernest Marsden (1889-1970) to carry out the experiment on the scattering of the alpha rays by the thin gold foil to test Thomson's Plum Pudding Model of an Atom. He had a source of alpha particle beam a ray through a circular detector (a fluorescent screen) with an opening for the ray to hit the gold foil (Figure 1.5).

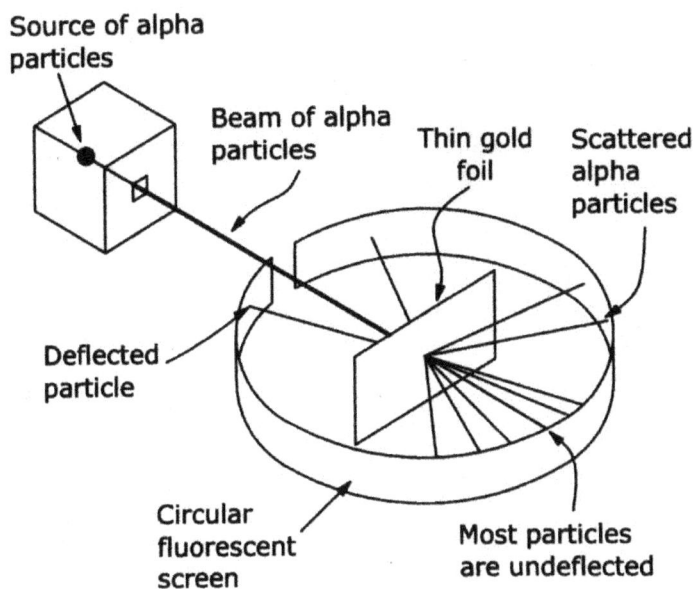

Figure 1.5. Experiment of alpha rays on a thin gold foil.

Rutherford reasoned that if Thomson's Model of an Atom was correct, the alpha particles should crash through the thin foil with minor deflections. The result showed that most of the alpha particles passed straight through, while many of the particles were reflected at large angles. But what greatly stunned Rutherford was that some particles were reflected, never hitting the detector, which means that the alpha particles were reflected almost back to the source.

What Rutherford made sense of this was that the Thomson's Model of an Atom was not right. In 1911, Rutherford proposed a new model of an atom; he suggested that negative charge electrons were distributed about a positive charge nucleus (Figure 1.6).

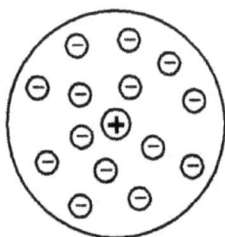

Figure 1.6. Rutherford Model of an Atom. Negative charge electrons were distributed about the positive charge nucleus.

Rutherford also discovered and coined the term "half-life" to describe the property of radioactive decay. In 1908, he received the Nobel Prize in Chemistry for his work on radioactivity.

Bohr's Model of an Atom (1913): Niels Bohr

In 1900, Max Planck (1858-1947), a German physicist created the Quantum Theory to explain the blackbody radiation.

Niels Bohr (1885-1962), a Danish physicist had earlier studied Planck's Quantum Theory for his doctoral dissertation. After he obtained his doctorate in 1911, he worked for Rutherford for a while.

In 1913, Bohr presented his model of an atom (Figure 1.7).

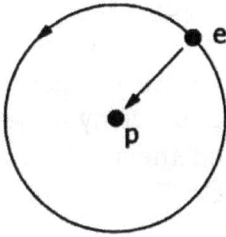

Figure 1.7. Bohr's Model of an Atom. The model is that of a hydrogen atom where an electron is in circular orbit around a single proton. The centripetal force (the forced directed or moving toward the center or axis) is provided by the Coulomb attraction (the electrostatic force) between the electron and the nucleus.

Bohr forwarded two postulates for his model:

- The electron moves only in a certain circular orbits, called stationary states.
- Radiation occurs only when an electron goes from one allowed orbit to another of lower orbit.

The postulates does not explain why the atom is stable; it merely asserts that it is.[14] From these conditions, Bohr was able to explain the emission spectrum of hydrogen.

The Discovery of Proton (1919): Ernest Rutherford

The concept of a positive charge particle that resides in the nucleus of an atom had earlier been suggested by the works of Eugen Goldstein.

Rutherford's experiment led to the discovery of the proton. In 1917, Rutherford became the first scientist to split the atom. Rutherford fired the alpha particles into a container filled with nitrogen gas with the result monitored through a fluorescent screen behind it. The alpha particles are normally absorbed by the nitrogen but sometimes flashes of new particles more penetrating than the alpha particles were emitted. Rutherford found that the nitrogen struck by alpha particles had changed into oxygen, while releasing at the same time the nucleus of hydrogen. Rutherford speculated that the nucleus of hydrogen was the building blocks of all nuclei. He called the nucleus of the hydrogen as "proton," meaning "first particle."[15]

The Discovery of Neutron (1932): James Chadwick

In 1920, Ernest Rutherford suggested the existence of a "neutral element" within the atom and was proposed later by Santiago Antúnez de Mayolo (1887-1967) at the Third Scientific Pan-American Congress in 1924.[16]

James Chadwick (1891-1974), a British physicist discovered neutron in 1932.

This completes the fundamental particles that made up the atom: electron, proton, and neutron. The discovery of the neutron is important as it made it possible to penetrate and split the nuclei of even the heaviest elements in what is called *nuclear fission*. For his discovery, Chadwick was awarded the Hughes Medal of Royal Society in 1932 and the Nobel Prize for Physics in 1935.[17]

The Atom

Our understanding of the atom is usually abstracted by the world around us as we normally don't think of an atom in our everyday life unless we are studying about it. It is for this reason that we cannot imagine the relative dimension of the subatomic particles inside the atom such as the distance between the nucleus and the electron and the enormous amount of atoms that made up a single period in this page. Understanding the world of subatomic particles would give us an idea of the world we live in and nature of the things around us.

Dimension and Scale of an Atom

The size of the atom depends on the element. The typical nucleus of the atom has a diameter of about 10^{-14} meter, while the diameter of the whole atom is about 10^{-10} meter. To imagine the dimension of an atom, the following are the comparison of the diameter of the nucleus to the whole atom:

- If the nucleus is the size of a pea, the electron would be at a distance of 1/4 mile (400 meters) from the nucleus.[18]
- If the nucleus is the size of a golf ball, the electron would be about 0.6 mile (1 kilometer).[19]
- If the nucleus is the size of basketball, the electron would be about 20 miles (32 kilometers) away.[20]

As it is, the atom is consists mostly of empty space.

Another way to imagine how small the atom is, a typical period in this page which is made of carbon atoms will have 5×10^{18} atoms.[21]

Mass and Charge of Proton, Neutron, and Electron

Common depictions of the proton, electron, and neutron practically cannot show their relative sizes but this can be inferred from the relative mass of the particles (Table 1.1).

Lifetime of Proton and Neutron

Free neutron was found to decay (changes to proton, and emits electron and electron antineutrino) in about 13 minutes. Proton, on the other hand, is considered as a stable particle as spontaneous decay of free proton has never been observed. However, various experiments such as in Super-Kamiokande detector in Japan gave a lower limit of the lifetime of the proton as 10^{33} years, while that of the Sudbury Neutrino Observatory in Canada put the lifetime of the proton at 10^{29} years. [22]

Table 1.1 Properties of proton, neutron, and electron

Subatomic particle	Charge*	Mass (kg)	Relative mass**
Electron	-e	9.109×10^{-31}	1
Proton	+e	1.673×10^{-27}	1,836
Neutron	0	1.675×10^{-27}	1,839

* The magnitude of charge of the electron and proton is 1.60×10^{-19} C.
** Relative to electron's mass.

Electron Spin

Classically, electron is pictured physically as a charge sphere that is spinning on its axis (Figure 1.8).

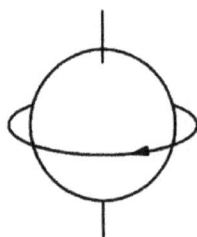

Figure 1.8. Electron pictured as a spinning sphere.

There are two observations that can account for the spin of an electron:

- In iron and other ferromagnetic materials, the spin of the electron accounts for the strong magnetic results.

- In 1924, Wolfgang Pauli proposed the existence of a fourth quantum number to provide solution to the problem of the observation of the so-called fine structure present in many spectral lines, where for example, a spectral line of sodium when examined under high resolution was found to consists of two lines. Erwin Schrödinger's Wave Equation cannot account for the fine structure nor can it predict correctly the number of new lines that appears when the atom was placed in a magnetic field in what is called the *Zeeman Effect*. The proposal of Pauli led S.A. Goudsmit (1902-1978) and E. Uhlenbeck (1900-1988) in 1925 to suggest that each electron has an intrinsic spin angular momentum, *z*, where they pictured the electron as a charge particle spinning about an internal axis. Pauli worked out the theory in depth in 1927 and the following year Paul Dirac derived his Relativistic Quantum Mechanics, where the electron spin was an essential part of it.

Ironically, according to Quantum Mechanics and particle physics, spin is solely a "quantum-mechanical phenomenon" and it does not have a counterpart in the Classical Mechanics.[23]

(The physical spin of a charged particle is the cornerstone of my *Charge Theory* in Chapter 3.)

What Holds the Electron in Orbit Around the Nucleus

Why the electron orbits around the nucleus is not very much clear in physics. In Bohr's Model of an Atom, the electron orbiting the nucleus is like the moon orbiting the Earth or the Earth orbiting the Sun. The centripetal force acting between the orbiting body and the center is explained as the electrostatic force or Coulomb attraction between the electron and the nucleus.

What Holds the Nucleus of the Atom Together

Physicists are mystified by the structure of the nucleus of the atom. They knew that for the element higher than hydrogen, the nucleus is composed of protons and neutrons. As it is, there should be a problem since the repulsion between the protons in the nucleus should be so great that the nucleus should explode. Physicists called the force holding the nucleus together as the strong nuclear force or strong interaction (which is one of the four fundamental forces of nature that will be discussed in Chapter 2).

The Structure of an Atom

In the discussions on the structure of an atom, there are ironically co-existing two models of an atom that are seemingly at odds with each other, yet not often recognized. Bohr's Model of an Atom shows a definite orbit of an electron around the nucleus, while that of Schrödinger's Model of an Atom shows a random revolution of an electron in a region called *electron cloud*. Due to the random movement of the electron in the electron cloud, the term "orbital," coined by Robert Millikan in 1932, was used to differentiate it from the term "orbit."

Pauli Exclusion Principle (1925): Wolfgang Pauli

In 1925, Wolfgang Pauli (1900-1958), an Austrian theoretical physicist in a study of the classification of the spectral lines led him to formulate what is known as the *Pauli Exclusion Principle* as a way to explain the arrangement of electrons in an atom. The Pauli Exclusion Principle also provided a theoretical basis for the modern Periodic Table. Pauli was awarded the Nobel Prize in Physics in 1945 for his principle.

The Pauli Exclusion Principle states that:

No two electrons in an atom can have the same four quantum numbers n, ℓ, m_ℓ, and m_s .

The following are the descriptions of the quantum numbers for an electron:

- Principal quantum number (*n*). The electron shell or energy level of an atom. The value of *n* is from the first energy shell 1 to the outermost electron shell of atom (corresponding to the labels: *K*=1, *L*, *M*, *N*, *O*, ...).
- Orbital (*ℓ*). The subshell or sub-energy level of the electron of an atom. Also known as angular or azimuthal quantum number. The value of ℓ could be from 0, 1, 2, 3, ... (corresponding to the labels: *s*, *p*, *d*, *f*, ...).
- Magnetic (*m_ℓ*). Describes the specific orbital (electron cloud) within the subshell.
- Spin (*m_s*). Describes the spin of the electron within that orbital. The values of spin could only be $m_s = \pm 1/2$. The spin +1/2 of the electron means that it is "spin up" or spinning counterclockwise when looking down, while the spin -1/2 means that it is "spin down" or spinning clockwise when looking down at the electron.

For atoms with more than one electron, the energy levels depend on both *n* and ℓ. That is, Pauli Exclusion Principle is not applicable to the hydrogen atom, which has only one electron.

Shell Model of the Electron of an Atom

The British physicists Charles Barkla (1877-1944) and Henry Moseley's experiments on the study of X-ray absorption provided the first observation of the existence of electron shells. Barkla labeled the shells with letters *K*, *L*, *M*, *N*, *O*, *P*, and *Q*. Likewise, the subshells were labeled with the letters *s*, *p*, *d*, and *f* referring to the first letter of "sharp," "principal," "diffuse," and "fundamental," which were historical relics of the experimental terms referring to the spectra.

The maximum electron for the subshell is shown in Table 1.2 and the maximum electron for the shells is shown in Table 1.3.

The list of the atom's occupied energy levels is called its *electron configuration*. Figure 1.9 shows a useful mnemonics for the filling order atomic orbitals.

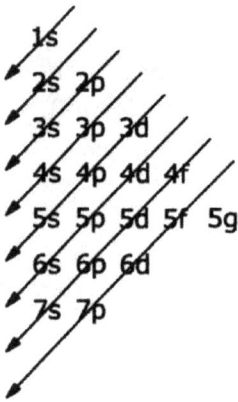

Figure 1.9. Mnemonics for filling order of atomic orbitals.

Based from filling order, an electron configuration pattern is as follows:

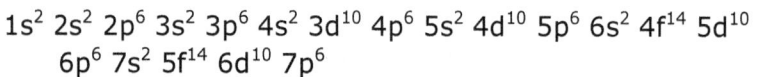

$$1s^2\ 2s^2\ 2p^6\ 3s^2\ 3p^6\ 4s^2\ 3d^{10}\ 4p^6\ 5s^2\ 4d^{10}\ 5p^6\ 6s^2\ 4f^{14}\ 5d^{10}$$
$$6p^6\ 7s^2\ 5f^{14}\ 6d^{10}\ 7p^6$$

Table 1.2 Maximum capacity of the subshell

Subshell	Maximum capacity
s	2
p	6
d	10
f	14
g	18

Table 1.3 Maximum electrons in a shell

Shell	Subshell	Subshell Max. electrons	Shell Max. electrons*
K	1s	2	2
L	2s	2	8
	2p	6	
M	3s	2	18
	3p	6	
	3d	10	
N	4s	2	32
	4p	6	
	4d	10	
	4f	14	

* Total number of electrons in the subshells.

The Schrödinger Equation (1926): Erwin Schrödinger

In 1909, Albert Einstein had shown that a complete description of cavity radiation requires both the particle and wave aspect of radiation giving rise to the wave and particle property of light (photon)—the Wave-Particle Duality.

In 1924, Louis de Broglie (1892-1987), a French physicist hypothesized that a similar Wave-Particle Duality might apply to particles of matter. That is, matter may also display a wave behavior.

When Erwin Schrödinger (1887-1961), an Austrian physicist heard de Broglie's hypothesis, he did not believe it at first but since Einstein supported de Broglie's idea, Schrödinger decided to look for an equation to describe the waves of matter.

In 1926, Schrödinger forwarded the resulting equation called the *Schrödinger Equation*, which is a complex mathematical equation describing the changing wave pattern of a particle such as an electron in an atom. The solution of the equation gives the probability of finding the particle at a particular place.

Schrödinger believed that a particle is a group of waves, a wave packet, somewhat like a fuzzy powder puff. The implication of Schrödinger Equation is that it is no longer possible to detect exactly where a single particle, such as an electron in an atom, will be detected.

In 1933, Schrödinger was awarded the Nobel Prize in Physics for his work on wave mechanics.

Schrödinger's Model of an Atom: Electron Clouds and Orbitals

In Schrödinger's Model of an Atom, the electrons are arranged in **orbitals**, which were systematically distributed within **electron clouds** (Figure 1.10). Schrödinger defined an orbital as the region of space that surrounds a nucleus in which *two* electrons may randomly move.[24]

Note: The two electrons in the orbital is in accordance with Pauli Exclusion Principle where the electron can only have a value of spin down and spin up.

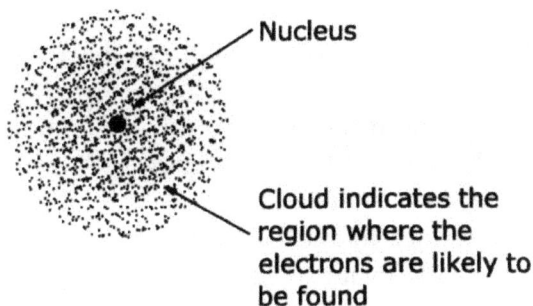

Figure 1.10. The electron cloud.

Schrödinger's Cat

In 1935, Schrödinger devised a "thought experiment" (of which Einstein was famous of using) to illustrate the probability of finding, say an electron, at a particular place. The thought experiment became famously called the *Schrödinger's Cat*. The idea of the experiment is that of a close box containing a radioactive material, a canister of cyanide (a poison), a detector mechanism connected to a hammer, and a live cat. In the setup, the ejection of a particle by the radioactive material would trigger the detector to break the canister of the cyanide that would kill the cat. Schrödinger argued that with no way of knowing what is going on inside the box, the cat could either be dead or alive until the box is opened. The question is: Would the cat be alive or dead before opening the box? (It is said that this paradox has not yet been fully resolved.)

The Heisenberg Uncertainty Principle (1927): Werner Heisenberg

In 1927, Werner Heisenberg (1901-1976) developed the *Heisenberg Uncertainty Principle*, a different form of Quantum Mechanics that was later shown to be equivalent to that of Schrödinger. The Heisenberg Uncertainty Principle stated that it is not possible to determine both the position and momentum of a particle (such as an electron) simultaneously to an arbitrary precision.

For example, to measure both the position and momentum of a particle simultaneously requires two measurements, which in the process of doing the first measurement will disturb the particle, creating an uncertainty in the second measurement. According to Heisenberg, this inability to get an accurate result in the experiment has nothing to do with experimental skill or equipment, rather, it is a fundamental restriction imposed on us by nature.

Heisenberg was awarded the 1932 Nobel Prize in Physics for his discovery.

Nuclear Shell Model of an Atom (1949): Maria Goeppert-Mayer

The Nuclear Shell Model is the model of the nucleus of an atom that uses Pauli Exclusion Principle to describe the structure of the nucleus in terms of energy levels. The first shell model was proposed by Dmitry Ivanenko (1904-1994) together with E. Gapon in 1932.[25]

Maria Goeppert-Mayer (1906-1972), a German-born physicist studied theoretical physics in Germany before coming to the United States in 1939 with her husband Joseph Mayer (1904-1983), an American chemist. In 1946, she and her husband moved to Chicago where she worked half the time in the University of Chicago and Argonne National Laboratory. She worked under Edward Teller (1908-2003) on a project to determine the origin of the elements. Her work involved creating a list of isotope abundancies where it became clear to her that the nuclei 2, 8, 20, 28, 50, 82, or 126 protons or neutrons called "magic numbers" were especially stable. This observation led her to suggest a shell model for nuclei similar to the electron shell model of an atom.

While collecting data in support of her nuclear shell model, she was having difficulty in her theoretical explanation. The answer came during her discussion with Enrico Fermi (1901-1954). Fermi casually asked her if there was an evidence of spin-orbit coupling. Spin-orbit coupling happens when two motions are coupled together, such as the Earth spinning on its axis as it orbits the Sun—which in an atom, the electron spins on its axis as it orbits the nucleus.[26]

As she was sending her paper for publication in the *Physical Review*, she became aware of a paper by Hans Jensen (1907-1973) and his colleagues, who had independently came up with the same result. She asked that her paper be delayed so she could publish it in the same issue as theirs, but hers ended up being published in the issue after theirs in June 1949.

Goeppert-Mayer won the Nobel Prize in Physics in 1963, which she shared with Hans Jensen and Eugene Paul Wigner (1902-1995).

Orbital Diagram: Aufbau Principle, Pauli Exclusion Principle, and Hund's Rule

Another way to represent the electron configuration is by using an *orbital diagram* consisting of boxes representing orbitals and arrows, which indicate the state of an electron (Figure 1.11).

The three rules used in filling the orbital diagram of an atom are: Aufbau Principle, Pauli Exclusion Principle, and Hund's Rule.

- The Aufbau Principle (Aufbau means "building up" in German) is stated as follows: Orbitals are filled beginning from the lowest energy to the highest energy.
- The Pauli Exclusion Principle stated as: *No two electrons in an atom can have the same four quantum numbers*) can be translated in terms of the orbitals as: *In an orbital which can have two maximum electrons, no two electrons can have identical spin*. That is, an orbital that

has two electrons, one should be a "spin up" and the other a "spin down" electron.

- The Hund's Rule, the third principle, states that: When filling the subshells other than the s subshell, electrons are placed in individual orbitals before they are paired up.

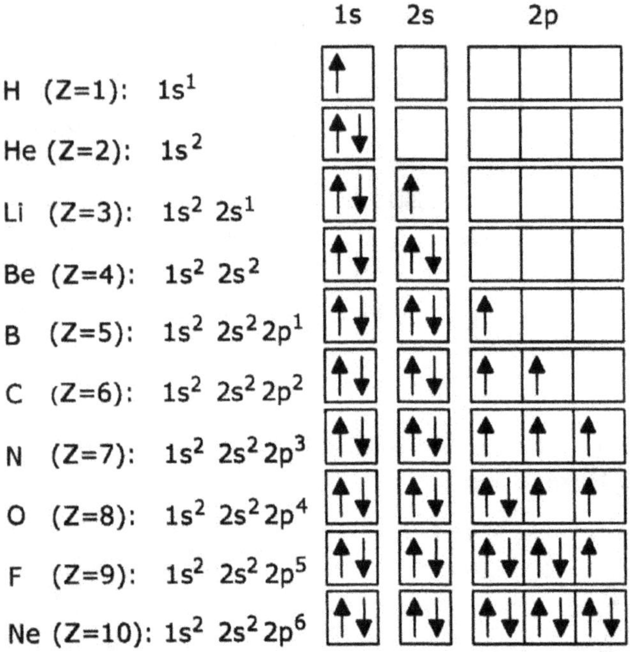

		1s	2s	2p
H	(Z=1): $1s^1$	↑		
He	(Z=2): $1s^2$	↑↓		
Li	(Z=3): $1s^2\,2s^1$	↑↓	↑	
Be	(Z=4): $1s^2\,2s^2$	↑↓	↑↓	
B	(Z=5): $1s^2\,2s^2 2p^1$	↑↓	↑↓	↑
C	(Z=6): $1s^2\,2s^2 2p^2$	↑↓	↑↓	↑ ↑
N	(Z=7): $1s^2\,2s^2 2p^3$	↑↓	↑↓	↑ ↑ ↑
O	(Z=8): $1s^2\,2s^2 2p^4$	↑↓	↑↓	↑↓ ↑ ↑
F	(Z=9): $1s^2\,2s^2 2p^5$	↑↓	↑↓	↑↓ ↑↓ ↑
Ne	(Z=10): $1s^2\,2s^2 2p^6$	↑↓	↑↓	↑↓ ↑↓ ↑↓

Figure 1.11. The orbital diagram shows the hydrogen (H), beryllium (Be), lithium (Li), boron (B), carbon (C), nitrogen (N), oxygen (O), fluorine (F), and neon (Ne) atoms are filled with electrons based on the electron configuration of an atom (Z=Number of electrons of an atom).

From Figure 1.10, the Aufbau Principle is shown by the filling of the orbital diagram of the subshell starting from *1s, 2s, 2p...* From the Pauli Exclusion Principle, each orbital diagram box contains only two electrons with spins of spin up and spin down. The Hund's Rule is shown from the subshell *2p* of the Boron (B) to the Neon (Ne) where each box is first filled with a single spin up electron until the three boxes are filled before going back to the first box where a spin down electron is added.

Logo of an Atom

The depiction of an atom in a logo follows the developments in physics. While Schrödinger's Model of an Atom shows the random movement of the electrons around the nucleus, the image of the logo of an atom requires the need for a well-proportioned and organized image to represent the atom. One of the famous depictions of an atom was the seal of the defunct US Atomic Energy Commission (AEC) whose atom in its logo was incorporated in the seal of the US Department of Energy (DOE) (Figure 1.12).

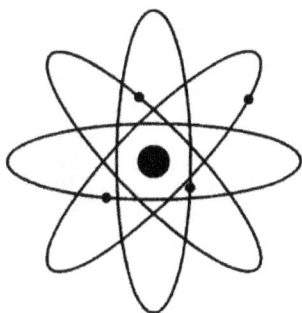

Figure 1.12. Depiction of the atom in the US Department of Energy (DOE) seal showing four orbiting electrons and a representation of the nucleus with a single particle.

The lithium atom presents a more aesthetic representation of the atom with its three orbitals, which are actually made up of two shells (Figure 1.13).

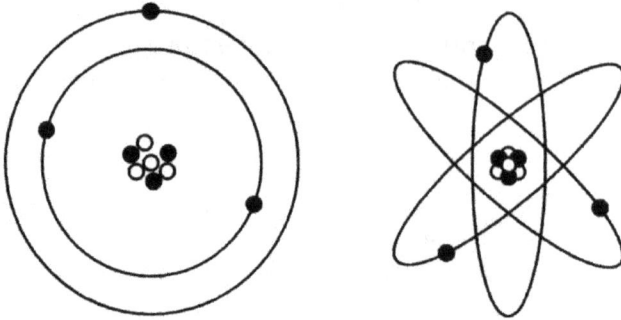

Figure 1.13. The lithium atom with its electrons on its shells and the nucleus lumping four neutrons and three protons.

A more common depiction of an atom is that of the three-orbiting electrons with a single particle as a nucleus (Figure 1.14).

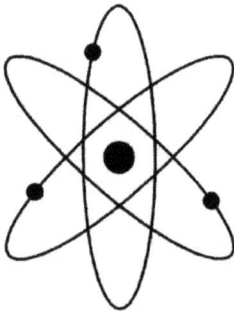

Figure 1.14. A simplified and stylized depiction of an atom.

Bottom line is that the logo shows that of the random orbits (in the so-called orbitals) of the electrons and the disorganized lumping of the neutrons and protons as the nuclei.

Note

There are two main questions that should be pointed out in this chapter: one is on the structure of an atom by the model of an atom and the other is on the subject of spin of the particles in physics.

Model of an Atom

When I wrote my first book on physics, my theory on light, I did the same of describing the early developments in the model of an atom: from J.J. Thomson's Plum Pudding Model to Hantaro Nagaoka's Saturn Model of an Atom to Ernest Rutherford's Model of an Atom to Niels Bohr's Model of an Atom. Nowhere did I read from the books the Schrödinger's Model of an Atom. That is, the model of an atom passes from Bohr's Model of an Atom exhibiting the solar system-like model of an atom with its fixed orbit of the electron to the random movements of the electrons around the nucleus with its terms of "orbitals" and "electron cloud"—without much calling into attention on the significant changes on the structure of the atom.

As to the nucleus of the atom, there is also the Nuclear Shell Model of an Atom that does not show the actual physical arrangement of the protons and neutrons in the nucleus of the atom. Much of the illustrations of the nucleus of the atom are that of the protons and neutrons lump together as a nucleus. The nucleus of the atom is said to be bound by the strong nuclear force that prevents the repulsion between the protons in the nucleus that would otherwise set the protons flying off from the nucleus.

These are only some of the questions on the structure of the atom that is not given any attention. A very robust theory on the structure of an atom and a new model of an atom is needed to correct these misunderstandings. It should be a theory that should give us a very good understanding of the arrangement of the atom of the elements in the Periodic Table of the Elements and its characteristics, and it should also be robust to also answer any questions in chemistry.

Spin

As pointed out in the discussion on the section on *Electron Spin*, the spin of the electron passes from the actual spinning electron to being relegated only as a "quantum-mechanical phenomenon" that does not have a counterpart in Classical Mechanics.

* * *

The model of an atom had slipped into the unknown. With nobody to recognize that there is something wrong with the model of an atom, physics will keep marching on and everyone would think that everything is just okay with it. The only way to see the problem is to know what charge is, and to challenge the Bohr's Model of an Atom, Schrödinger's Model of an Atom, Pauli Exclusion Principle, and the Nuclear Shell Model of an Atom.

Chapter 2
The Standard Model

Part I: The Standard Model in a Nutshell

The Standard Model

The Standard Model is the name given to the theory developed throughout the mid to the late 20th century concerning the fundamental particles of matter and their interactions. The Standard Model is regarded sometimes as the Theory of *Almost Everything* since even with its success in explaining wide varieties of experimental results it cannot explain gravity.

Much like the Periodic Table of the Elements, the Standard Model is laid out in a table (Figure 2.1).

The Standard Model constitutes the building blocks of the universe. It is composed of twelve spin 1/2 fermions consisting of six quarks (up, down, charm, strange, top, and bottom) and six leptons (electron, electron neutrino, muon, muon neutrino, tau, tau neutrino), and four spin 1 gauge bosons (photon, Z boson, W boson). (Fermion was named after Enrico Fermi, while boson was named after Satyendra Nath Bose.) Notice that the hypothetical mediating particle of gravity (called *graviton*) is not included and also the Higgs boson that is a spin 0 boson.

Out of the six quarks, only two quarks occur in the ordinary matter: the *up* and *down* quarks that make up the proton and neutron. As such, it can be said that the universe is practically made of the *up* and *down* quark, which are generally more stable; and the lepton electron. Heavier quarks quickly change into *up* and *down* quarks through the process of particle decay. Table 2.1 shows the properties of the Standard Model.

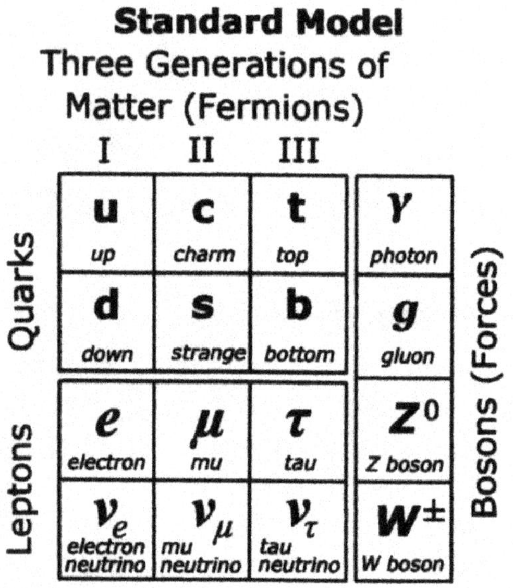

Standard Model
Three Generations of
Matter (Fermions)

	I	II	III	
Quarks	**u** *up*	**c** *charm*	**t** *top*	**γ** *photon*
	d *down*	**s** *strange*	**b** *bottom*	**g** *gluon*
Leptons	**e** *electron*	**μ** *mu*	**τ** *tau*	**Z⁰** *Z boson*
	$\boldsymbol{v_e}$ *electron neutrino*	$\boldsymbol{v_\mu}$ *mu neutrino*	$\boldsymbol{v_\tau}$ *tau neutrino*	**W±** *W boson*

(right side labeled: **Bosons (Forces)**)

Figure 2.1. The Standard Model.

The difference between a fermion and boson is that fermion has a half integer spin, while boson has an integer spin. Quarks are found in composites, while leptons are only found alone.

Quarks are formed into what is called hadron. Hadrons are further classified into meson and baryon. A baryon is made up of three quarks, while that of a meson is made up of a quark and an antiquark. (An antiquark notation is a quark denoted with a "bar" on top such as $\bar{u}$, which is read as bar-u.) Table 2.2 shows the list of baryons and mesons and their quarks.

Table 2.1 Properties of the fundamental particles and forces

Fundamental particle	Symbol	Mass*	Charge (e)	Spin
Quarks				
Up	u	2.4 MeV	+2/3	1/2
Down	d	4.8 MeV	-1/3	1/2
Charm	c	1.27 GeV	+2/3	1/2
Strange	s	104 MeV	-1/3	1/2
Top	t	171.2 GeV	+2/3	1/2
Bottom	b	4.2 GeV	-1/3	1/2
Leptons				
Electron	e	0.511 MeV	-1	1/2
Muon	μ	105.7 MeV	-1	1/2
Tau	τ	1.777 GeV	-1	1/2
Lepton Neutrinos				
Electron neutrino	ν_e	0 [$<4 \times 10^{-6}$ MeV]	0	1/2
Muon neutrino	ν_μ	0 [<0.17 MeV]	0	1/2
Tau neutrino	ν_τ	0 [<18 MeV]	0	1/2
Bosons				
Photon	γ	0	0	1
Gluon	g	0	0	1
W	$W^\pm$	80.4 GeV	±1	1
Z	Z^0	91.2 GeV	0	1

* The 0 value for the neutrino is its rest mass, while the value in the square bracket is the experimental upper limit of its mass.

Table 2.2 List of some hadrons and its quarks

Particle*	Quark content
Baryons (Spin 1/2)	
Proton	uud
Neutron	udd
Λ^0	uds
Σ^+	uus
Σ^0	uds
Σ^-	dds
Ξ^0	uss
Ξ^-	dss
Ω^-	sss
Mesons (Spin 0)	
π^0	$u\bar{u}$, $d\bar{d}$
π^+	$u\bar{d}$
π^-	$\bar{u}d$
K^+	$u\bar{s}$
K^-	$\bar{u}s$

* Superscript of the particle indicates its charge.

Further Classification of the Particles

Figure 2.3 shows the fundamental particles of nature and how they were formed to what we experience as matter.

The only stable baryon is proton, while neutron decays into proton, electron, and electron antineutrino. Mesons have spin 0 or 1 and therefore are all bosons (Young and Freedman, p.1684).

Matter
|
Fermions
Quarks   Leptons
|
Hadrons

Mesons   Baryons
                (Nuclei)

Atom
|
Molecules

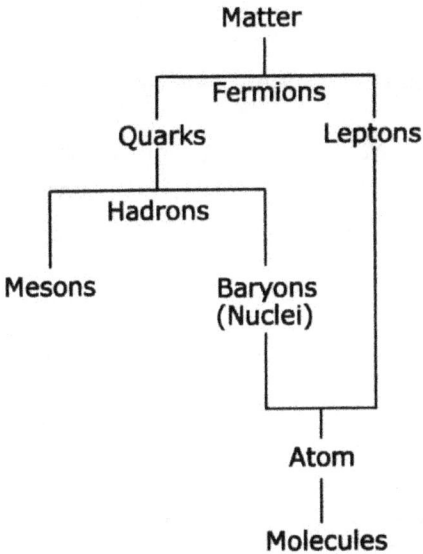

Figure 2.3. Diagram of the fundamental particles.

The Fundamental Forces

Fundamental forces (also called fundamental interactions) are the ways that the fundamental particles in the universe interact with one another. The four known fundamental forces are the following:

- Electromagnetism. The force that exists between all charge particles. Weaker than the strong force but long range.
- Gravitation. The force that holds us on Earth. The force that keeps the moon in orbit around the Earth and the planets around the Sun. Weak but very long range.
- Strong force. The force that is responsible for holding the nucleus of atom together. Strong but very short range.

- Weak interaction. The force that is responsible for nuclear beta decay and other similar decay processes involving fundamental particles. Weak and short range.

Figure 2.4 shows the diagram of the fundamental forces of nature, their mediating particle(s), the existing theory or theories being search for the individual forces, and the unification of two or more forces.

The theories that are still being search are Quantum Gravity Theory, Grand Unified Theory (GUT), and the Theory of Everything.

Table 2.3 shows the properties of the fundamental forces.

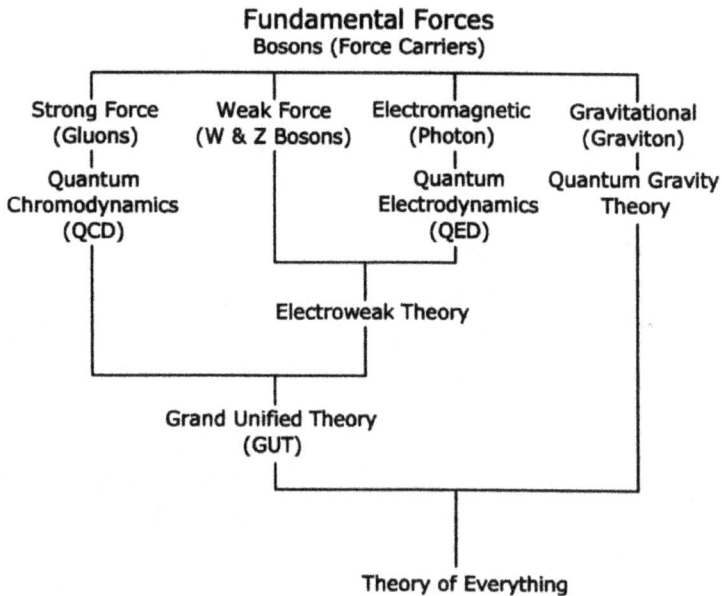

Figure 2.4. Diagram of the particles that mediates the force and their theories.

Table 2.3 Fundamental forces

Interaction	Mediating particle	Range (fm*)
Strong	Gluon	~1
Weak	$W^\pm$ and Z^0	~0.001
Electromagnetic	Photon	Infinite
Gravity	Graviton (hypothetical)	Infinite

* femtometer (fm) = 10^{-15} m.

Depiction of Quarks Inside the Proton and Neutron

Quarks are not actually observed outside of hadrons such as proton and neutron, hence have no known structure. They are typically depicted in various arrangements (Figure 2.2).

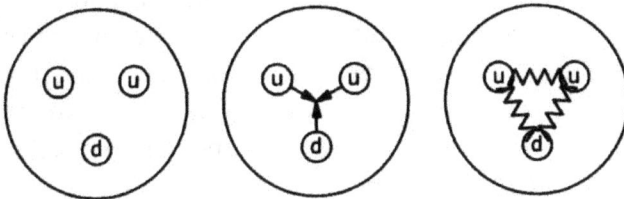

Figure 2.2. Depictions of quarks inside the proton. The neutron will have two *down* quarks and one *up* quark.

Theories and the Unification of the Forces

James Clerk Maxwell was said to be the first to unify electricity and magnetism in his Electromagnetic Theory. Albert Einstein tried to unify gravity and electromagnetism but failed. In 1960, Sheldon Glashow (b. 1932), Steven Weinberg (b. 1933), and Abdus Salam (1926-1996) independently worked out a theory that unified electromagnetism with weak interaction into

Electroweak Theory. (The weak interaction or weak nuclear force refers to the beta decay or to the radioactive decay of neutron.)

Physicists are still trying to unify all the forces today with either a single theory or unifying two or more forces in a piecemeal approach aiming for that Theory of Everything.

The Search for the Source of Mass of the Particles

In the theory of the Standard Model, bosons are supposed to be massless particles such as shown by the photon and gluon (refer to Table 2.1). The problem occurred when the W and Z bosons were found in the experiments to have a mass that were quite heavy. This gave rise to a theory needed to give mass to the W and Z bosons.

In 1964, Robert Brout (1928-2011), François Englert (b.1932), Peter Higgs (b.1929), Gerald Guralnik (b.1936), C.R. Hagen (b.1937), and Tom Kibble (b.1932) proposed a mechanism that would cause the particles to acquire mass. The Higgs Mechanism called the *Higgs field* could be proven by searching for the particle that it gives off. Just like the electromagnetic field with a photon as the mediating particle, the Higgs field mediating particle is called the Higgs boson (which is spin 0).

In 1967, Steven Weinberg (b.1933) and Abdus Salam (1926-1996) incorporated the Higgs Mechanism into Glashow's Electroweak Theory to take into account the absence of mass of the particle in the theory.

Challenges of the Standard Model

The following are just some of the challenges on the Standard Model which can be divided into two categories: one is pertaining to the mass of the particles and the other relates to cosmology:

44

- Neutrino mass. The Standard Model does not allow masses for the neutrino in spite of experimental evidence that neutrinos do have a mass.
- Higgs boson. The Standard Model does not account for the mass of the particles. To explain for the source of mass of the particles, the Higgs Mechanism called the Higgs field was invented, which has a mediating particle called the Higgs boson, to provide for the mass for the particles.
- Gravitation. The Standard Model cannot explain gravitation. Being a class of Quantum Field Theory, the Standard Model cannot be reconciled with the currently accepted theory on gravity, the General Theory of Relativity.
- Dark matter and dark energy. The Standard Model cannot explain the observed amount of dark matter that is responsible for preventing the galaxies from flying apart, as well as, the dark energy that is responsible for the observed acceleration of the universe.
- Matter/Anti-matter symmetry. The Standard Model cannot explain the predominance of matter over anti-matter in the universe.

The Standard Model may have been misunderstood by the physicists as they try to answer the various phenomena that they had observed or encountered in their theories and experiments. The Standard Model is both a theory and an empirical fact. It is an empirical fact in the sense that its constituent particles and forces were observed in nature and proven from the experiments to exist. It's a theory in that as a source of fundamental of particles and forces, it should be able to be used to explain the various phenomena in nature that are observed.

In science, observation always trumps a theory. In the case of the mass of particles, if a theory says that a particle has no mass and it is observed that it has a mass, then the question is on the theory and not on the observed fact. While there was an early opposition of the physicists to the Higgs Mechanism, the greatest mistake of the physicists (brought about by having no other better theory or understanding to explain the source of mass of the particles) is to let in the idea of the existence of the Higgs boson.

Part II: The Way to the Standard Model

The Saga of the Standard Model

The development of the Standard Model is said to be the crowning achievement of mankind. From the acceptance of the reality of the atom to the discovery of its subatomic particles, science had moved beyond to the discovery of the building blocks of our world and the forces that governed it. Men and women of science discovered, theorized, and probed the atom, coaxing it of its secrets. This is the saga of the Standard Model.

The Discovery of X-ray (1895): Wilhelm Conrad Röntgen

In the 1850s, physicists discovered and experimented with the cathode rays from the cathode ray tube (see Chapter 1 on *The Discovery of Electron (1897): J.J. Thomson*).

In 1892, Philipp Lenard (1862-1947), a German physicist found that the rays could pass through a thin aluminum window in the tube.

In 1895, Wilhelm Conrad Röntgen (1845-1923), a German physicist having learned of Lenard's experiment tried to see it for himself. He covered the cathode ray tube with a black paper, darkened the room, and switched on his apparatus. To Röntgen's surprise, he noted that at two meters away the fluorescent screen glows. He suspected a new kind of radiation more penetrating than the cathode rays. Unable to explain what the rays were, he called them X-rays ("X" for the unknown).

For his Röntgen's discovery, he won the Nobel Prize in Physics in 1901.

The Discovery of Radioactivity (1896): Henri Becquerel

On January 20, 1896, just two weeks after Röntgen announced his discovery of X-rays, the French Academy of Science held a special meeting. Henri Becquerel (1852-1908), a French scientist who attended the meeting was fascinated and was intrigued with the X-rays. Becquerel had done some research on materials that became luminous after its exposure to sunlight. He discovered by accident that the material containing uranium, even without exposure to sunlight, emitted radiations. He called the radiation as "uranium rays."

In 1895, Marie Curie née Maria Salomea Sklodowska (1867-1934), who came from Poland to study chemistry in Paris, chose uranium for her doctoral research. (She was later married to Pierre Curie (1859-1906), a French physicist.)

The Curies wanted to know if other substances emitted rays. They discovered that the metal thorium also emitted rays. They also discovered that an ore called *pitchblende*, from which the uranium was extracted, even emitted more intense rays than the uranium itself. In 1898, they identified the two radioactive substances as polonium (named in honor of Marie Curie's home country Poland) and radium. They were the first to prove that the atom of certain elements are continually breaking down and emitting radiation in the process. They called these elements as "radioactive."

In 1903, Becquerel and the Curies shared the Nobel Prize for Physics. Marie Curie was the first person to win two Nobel Prize when she was awarded another Nobel Prize in Chemistry in 1906.

Unaware of the dangers of radiation, Marie Curie was still at work in isolating a new element when she died. After her death, her body tissues were found to have been poisoned by too much exposure from radiation.

Paul Dirac's Antimatter Theory (1928) and Carl Anderson's Discovery of Positron (1932)

Earlier in 1898, Arthur Schuster (1851-1934), a British physicist suggested an idea that an exotic type of matter could exist with properties that mirror those of ordinary matter.

In 1928, Paul Adrien Maurice Dirac (1902-1984), a British physicist (originally trained as an electrical engineer) incorporated relativity with the newly developed Quantum Theory. His theory seemed to imply that free electrons have a positive counterpart, that is, a particle having the same mass as electron but is positive charge.

In 1930, Robert Millikan (1868-1953), an American experimental physicist at Caltech instructed fellow American physicist Carl Anderson (1905-1991), one of his research assistants, to build a cloud chamber to study the energies of cosmic ray particles. The result from the cloud chamber showed an equal number of positive charge and negative charge particles. Millikan originally insisted that the positive charge particles must be protons. On the other hand, Anderson believed that the tracks could be due to electrons moving upwards through the chamber rather than positive charge particles moving downward. To settle the debate, Anderson inserted a lead plate across the chamber that would lose some of the energy of the particles and that it would curve it more due to the magnetic field upon emerging on the other side of the lead plate. The result showed both Millikan and Anderson were wrong. The positive charge particle showed that it is moving up the lead plate in a counterclockwise direction. In

48

1932, Anderson discovered the antielectron or "positron" that was predicted earlier by Dirac.

Neutrino Postulate (1931): Wolfgang Pauli

Experiments on beta decay of the nucleus, wherein the neutron turns into proton and emits an electron, do not seem to follow the Law of Conservation of Energy as there is a missing energy. This created a problem for the physicists.

Wolfgang Pauli (1900-1958) provided the solution to the problem. Unable to attend a physics meeting, Pauli instead sent a letter proposing a new particle with zero charge and zero mass that was released in such reactions to account for the "missing energy." The idea was first heard in public at a physics meeting in Pasadena, California in 1931.

Pauli initially called the new particle as "neutron." But in 1932, when asked if this new particle was the same as the neutron of Chadwick, Enrico Fermi replied that Pauli's "neutron" was very much smaller. Fermi called it the "neutrino," which is an Italian word for "little neutral one."

Pauli earlier believed that the new particle could never be observed directly and that the neutrino will remain a hypothesis. In 1956, Clyde Cowan (1919-1974) and Fred Reines (1918-1998), both American physicists from Los Alamos National Laboratory in New Mexico finally detected the existence of the neutrinos.

The Early Atom-Smashers

In 1911, Rutherford and his assistants fired alpha particles to a thin gold foil to discover that an atom has a positive nucleus and is mostly made up of empty space.

In 1914, Rutherford's assistant Ernest Marsden (1889-1970), an English-New Zealander physicist continued the previous experiment conducted in 1911. He noticed that when the alpha particles are in action, a fast-moving "H-particles" (as the nucleus of hydrogen was then called) are sometimes seen. Rutherford who joined the hunt for the H-particles thought they were an unknown light gas. With the World War I happening, Rutherford's research was hampered.

It wasn't until the end of the war in 1919 that Rutherford continued the investigation by firing alpha particles on the nuclei of nitrogen producing oxygen and emitting the nuclei of hydrogen (that led to the discovery of proton). It was the first time that anyone had seen the transmutation of the elements— the vindication perhaps of the long search of the alchemists.

The following years, physicists kept on firing alpha particles at the nuclei relying on the fluorescent screen to observe the image with no way of recording the results. It was only in 1925 when Patrick Blackett (1897-1974), an English physicist using his own cloud chamber design (discussed below) in Rutherford's laboratory in Cambridge that a visual record of nuclear disintegration was seen. In 1948, Blackett received the Nobel Prize in Physics for his work.

As the alpha particles were recognized to be inefficient projectile, Rutherford called for other artificial sources that would provide for greater energy nuclear "bullets." In 1924, John Cockcroft (1897-1967), a British physicist who joined Rutherford's Cambridge laboratory used protons that were accelerated by a high-voltage electric field. At first, the energy that was required to break a nucleus seemed too high. George Gamow (1904-1968), a young theoretical physicist and cosmologist from Leningrad who visited Cambridge claimed that a lower-energy protons could drill into the nuclei owing to the effect called "quantum tunneling." It was with this knowledge that in 1932 Cockcroft and his collaborator Ernest Walton (1903-1995), an Irish physicist achieved the first completely artificial transformation of the element—the first time the atom was split. Watson and Cockcroft did this by ripping out the proton from the hydrogen gas and accelerated it at 800,000 volts to hit the lithium target. Watson watch through the microscope as the lithium nuclei changes into helium.

Cosmic Ray Particles

With the discovery of radioactivity, physicists measured radiation from the elements using an instrument called *electroscope*. An electroscope is consists of two gold leaves that are set close to each other and that when given an electric charge repeal each other. As the emissions from the radioactive material knocks the electrons from the atoms in the air, the air became ionized, making the air slightly conducting. When the electroscope was placed in the ionized air, the leaves slowly attract each other as the charge leaks away. It was in this experiment set-up it was discovered that the air seemed slightly conductive even when there was no radioactive material around. In 1900, this effect was reported by Charles Thomas Rees Wilson (1869-1959), a Scottish physicist. At first it was thought that the radioactive phenomenon came from the Earth.

In 1910, Theodor Wulf (1868-1946), a German physicist and a Jesuit priest using an electroscope compared the radiation from the ground to the top of the Eiffel Tower.

In 1911, Victor Hees (1883-1964), an Austrian physicist who read Wulf's work set off on a hydrogen balloon reaching to 5,350 meters and made careful measurements. He found out that as he went up, the ionization increases rapidly, with the ionization becoming several times more intense than the ground level as he reaches the 5,000 meters altitude. Hees concluded that the increase ionization had been caused by an unknown extraordinarily high or more penetrating radiation entering the atmosphere coming from the outer space. Hees won the Nobel Prize in Physics in 1936 for his work.

Cloud Chamber (1911): Charles Thomas Rees Wilson

In 1911, Charles Thomas Rees Wilson (1869-1959), a Scottish physicist invented the cloud chamber, which would become an important tool in the nuclear research and research on cosmic ray. (The term "cosmic rays" was coined by Robert Millikan, an American physicist, in 1925. He was also important in proving the Photoelectric Effect of Albert Einstein.)

Wilson who studied meteorology knew that the clouds were formed when the air saturated with water vapor cools and that condensation was triggered by the slightest impurity or disturbance. Wilson's cloud chamber reproduced the cloud formation using a set-up of a glass cylinder containing air and water vapor with a piston compressing the substances. When the piston was suddenly pulled out, the air in the cylinder caused it to expand and cool. (In thermodynamics, this is shown by the formula $P=VT$, where P is pressure, V is volume, and T is temperature.) In this set-up, when the particle passes through the chamber, the path formed from the condensation of the water droplets from the supersaturated water vapor showed the trail of the particle.

Positron (1931): Carl Anderson

In 1928, Paul Dirac predicted the existence of a positive counterpart of electron. In 1932, Carl Anderson discovered this antielectron particle called positron.

Muon (1936): Carl Anderson

In 1936, four years after his discovery of positron, Carl Anderson discovered muon during the study of cosmic rays using his magnetic cloud chamber. Muons are produced when particles from space, which are mostly protons, hit the atoms and nuclei in the upper atmosphere. Anderson noticed that some particles bend less than the electrons, which means that they must be heavier particle.

The collision actually creates unstable pions, which within much less a millionth of a second turns into muon and neutrino that shower down to the Earth's surface. Muon usually decays into an electron, electron-antineutrino, and muon neutrino.

The discovery of the muon was unexpected and did not fit with the atomic theories of that time, which prompted an American physicist I.I. Rabi to famously remarked, "Who ordered that?"

Pion (1947): Cecil Frank Powell and Group at Bristol

In 1935, Hideki Yukawa (1907-1981), a Japanese theoretical physicist proposed a (meson) particle that conveys the force between the protons and neutrons in the nucleus. Yukawa was building on an earlier Quantum Theory of electromagnetic forces developed earlier by Dirac in 1928.

In 1947, pion was discovered in emulsions exposed at Pic du Midi by the Powell Group from Bristol. In 1949, Yukawa received the Nobel Prize in Physics for his prediction of the pion.

Kaon (1947): George Rochester and Clifford Butler

In 1947, George Rochester (1908-2001) and Clifford Butler (1922-1999), both British physicists noticed an unusual effect of cosmic rays in their cloud chamber, where two tracks emerged from a point and forming an inverted V-shape. They concluded that an unknown particle decayed into two secondary particles.

Carl Anderson confirmed the discovery in 1950 from a cloud chamber photographs in White Mountain, California (Mt. Wilson). The new particle was called "K-meson" or kaon. Kaon is also dubbed as "strange particles" since it decay extremely slow at 10^{-10} seconds.

Bubble Chamber (1952): Donald Glaser

In 1952, Donald Glaser (1926-2013), an American physicist invented a new device called the bubble chamber to replace the cloud chamber. The cloud chamber was inefficient in detecting high-energy particles that just passes through the cloud vapor without leaving any tracks. The new device instead uses a superheated liquid that is just slightly above its boiling point but was kept under pressure to prevent it from boiling. Glaser received the Nobel Prize in Physics in 1960 for his invention of the bubble chamber.

Particle Accelerators and Detectors

Physics had come a long, long way on using charge particles to probe the atom and later smash it to get a thorough understanding of the atom and our world.

Ernest Rutherford used the alpha rays to discover the structure of the atom, that is, with a positive charge proton nucleus.

Cockcroft and Walton built a machine, the Cockcroft-Walton generator, using protons to propel it and bombard the atom and create a nuclear transformation.

The discovery of cosmic rays and the use of the cloud chamber and bubble chamber to record the results of the cosmic rays collisions produced a number of particles. The increasing number of particles led Murray Gell-Mann to forward the theory on quarks as part of the fundamental particles.

With the desire to know more about the atoms came the desire for more powerful machines that could smash the particles and supposedly give up more of its secrets. Modern particle accelerators use electric and magnetic field to propel and guide beams of charged particles in either a linear or circular path and to hit a fix target or to collide two beams.

Cyclotron

The cyclotron was invented by Ernest O. Lawrence (1901-1958) and Milton Stanley Livingston (1905-1986), both American physicists at the University of California, Berkeley in 1930. Cyclotron uses magnets to deflect the charged particles, usually protons, in a nearly circular path. The proton moves in a vacuum chamber in a uniform magnetic field that is perpendicular to the plane within the two D-shaped (called "dees") cavities (Figure 2.5).

Each time the proton passes into the gap between the "dees," the voltage accelerates them, increasing its speed and also the radius of curvature of its path. After many revolutions, the proton acquires high kinetic energy as it reaches the outer edge of the cyclotron. The proton then either strike a target or leave the cyclotron with the help of a carefully placed "bending magnet" and are directed to an external target. Lawrence's first cyclotron was just about 4 inches in diameter.

Synchrotron

A more practical machine to attain higher energies is called synchrotron. The synchrotron uses a series of electromagnets placed at the same radius from the center of the circle. The electromagnets are interrupted by gaps where high voltage accelerates the charged particles, usually protons.

The particles usually starts in a smaller accelerator that gives them a considerable initial energy before being injected to the synchrotron ring where they must move in a circle of constant radius, increasing the magnetic field slowly as they speed up. (This is the way the synchrotron deal with the relativistic increase in effective mass with speed, that is, by increasing the magnetic field in time as the speed increases.) Synchrotrons are typically enormous with radius of the ring in kilometers.

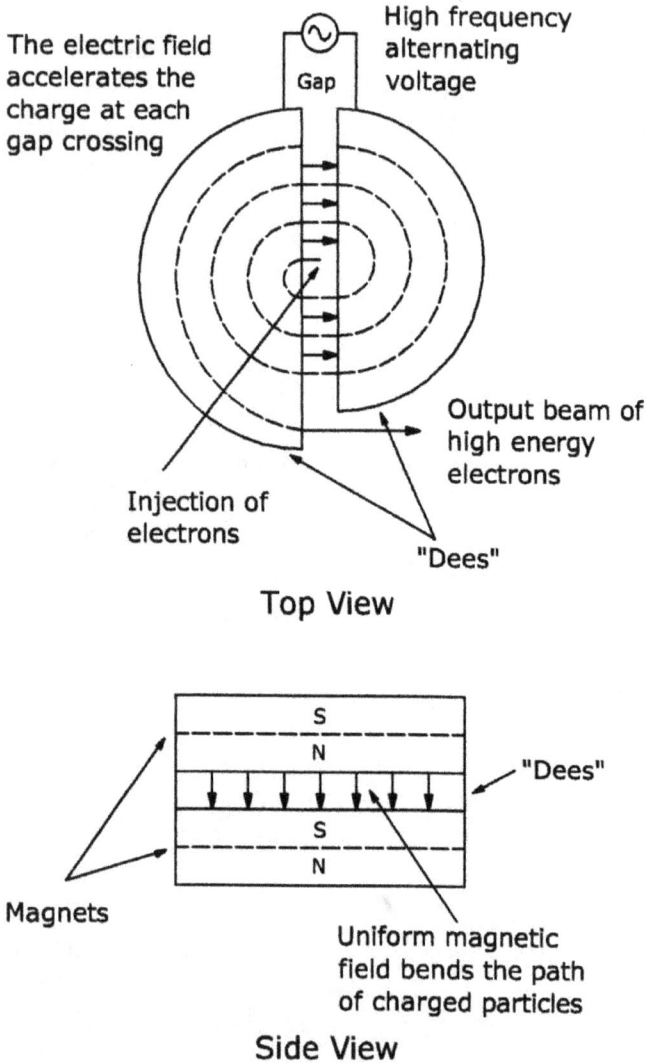

Figure 2.5. Schematic diagram of the cyclotron.

Linear Accelerator

A linear accelerator (linac) accelerates particles in a linear path. The linear accelerator is constructed of a series of tubular conductors with gaps to give the beam of particles a boost. The tubes are constructed longer the further they are from the source. Most present-day high-energy linacs use the travelling electromagnetic waves like the mag-lev train. Linear accelerators are typically used to accelerate electrons because their small mass makes them reach higher speed very quickly.

Colliders (Colliding Beams)

High-energy physics were originally carried out by beams of particle striking a stationary target. A very powerful technique called colliding beams, which obtains maximum possible collision energy from a given accelerator by accelerating at a very high energy two beams of particles and then steering them towards each other to collide head-on. This is accomplished using a single accelerator through the use of a storage ring connected to the main accelerator where one beam can continue to circulate, while the second beam is accelerated.

Detectors

Earlier detection of the particles used the cloud chamber and the bubble chamber. Modern detectors used in the high-energy particle accelerators uses wire chambers, which contained arrays of closely spaced wires that detect the ions. The information such as charge and time collected from each wire is processed using computers to reconstruct the particles trajectory.

The Big Machines

In particle physics, practice dictates an ever-increasing energy of the machines in order to smash the charge particles to discover more "hidden" particles. The available energy in the resulting head-on collision is two times the beam energy. There are a lot of particle accelerators around the world and the following are just the few of them:

Tevatron 1983-2011
Fermilab @ Batavia, Illinois
Synchrotron with energies up to 1 TeV, 6.28 km (3.9 mi)
Circular: Proton-Antiproton

Large Hadron Collider (LHC) 2008
CERN @ France and Switzerland near Geneva, Switzerland
Synchrotron with energies up to 7 TeV, 27 km (17 mi)
Circular: Proton-Proton

Stanford Linear Collider (SLC) 1989
SLAC @ Menlo Park, California, 50 GeV, 3.2km
Linear: Electron-Positron

The Race for the Discovery of Particles

The building of the big powerful machines naturally follows the race of finding the particles, both hypothesized and those that are still to be discovered.

Electron Neutrino (1956)

First hypothesized by Wolfgang Pauli in 1931 to account for the missing energy in the beta decay, the team led by Cylde Cowan (1919-1974) and Frederic Reines (1918-1998) discovered the electron neutrino in 1956.

Muon Neutrino (1962)

In 1962, muon neutrino was discovered by Leon Lederman (b. 1922), Melvin Schwartz (1932-2006), and Jack Steinberger (b. 1921). In 1988, they shared the Nobel Prize in Physics for the discovery.

Quarks (1964): Murray Gell-Mann and George Zweig

By the early 1960s, physicists had been inundated by well over 100 of hadrons discovered that they refer to them as "particle zoo." The fact that there were so many of them led the physicists to believed that they can't be fundamental particles. Their experiments indicated that the particles do have internal structure.

In 1961, Murray Gell-Mann (b. 1929), an American physicist at Caltech formulated a particle classification system known as the *Eightfold Way* (alluding to Buddhism's *Eightfold Path*) after he discovered how the different particles that were grouped into families containing eight or sometimes ten members fitted into special geometric patterns. Yuval Ne'eman (1925-2006), an Israeli physicist had also independently developed a similar scheme in the same year.

In 1962 at a physics conference, Gell-Mann predicted a new particle, the omega minus with its exact property, to fill the gap in a family of ten—in a manner reminiscent of Mendeleev's successful prediction of the elements in the gap in the Periodic Table. In 1963, a team at Brookhaven was able to detect one in their bubble chamber.

In the same year of 1963, Gell-Mann, and George Zweig (b. 1937) independently, proposed that hadrons were still not elementary particle but were actually composed of pointlike entities called "quarks." (Gell-Mann earlier used the word "quork" but upon reading James Joyce's book Finnegan's Wake, he came across the word "quark" in the line "Three quarks for Muster Mark!". Zweig chose the name "aces" but his ideas never got into print at the time.) At first, Gell-Mann was reluctant to risk his revolutionary idea that it will be refused by the cautious journal, so he submitted his paper to a relatively new European publication early in 1964.[1]

The model that Gell-Man and Zweig proposed involved three quarks they called "up," "down," and "strange." They also ascribed properties to the quarks such as spin and electrical charge. The initial reactions of the physics community to the proposed idea were mixed.

In less than a year, Sheldon Lee Glashow (b. 1932) and James Bjorken (b. 1934) predicted the existence of a fourth quark they called "charm."

Up and Down Quarks: Richard Taylor, Jerome Friedman, and Henry Kendall

In 1968, experiments at Stanford Linear Accelerator Center (SLAC) showed that the proton did contained much smaller, pointlike objects. The objects that were observed in SLAC would later be identified as the "up" and "down" quarks. The 1990 Nobel Prize in Physics for this discovery was awarded to Richard Taylor (b. 1929) from SLAC, and to Jerome Friedman (b. 1930) and Henry Kendall (1926-1999) from MIT.

Charm Quark and J/ψ Meson (1974): SLAC

The existence of the fourth quark (after up, down, and strange quarks) was speculated by a number of authors around 1964 but its prediction was usually credited to Sheldon Glashow, John Iliopoulos, and Luciano Maiani in 1970.

In 1974, the charm quark was found with the discovery of the J/ψ meson simultaneously by a team at the Stanford Linear Accelerator Center led by Burton Richter (b. 1931) and by the Brookhaven National Laboratory (BNL) led by Samuel Ting (b. 1936).

60

Worth mentioning perhaps to illustrate how competitive the race for finding the particles was how I learned about the simultaneous discovery of J/ψ meson by SLAC and BNL teams in Leon Lederman's book, *The God Particle*.[2] As the story goes, there was a controversy at who discovered it first. A charge on SLAC scientists was that they were aware of Ting's preliminary results and thus knew where to look accordingly. This was countered that Ting's initial result was inconclusive but was massaged in the time between SLAC's discovery and Ting's announcement. Having both claimed the particle for themselves; SLAC people named it ψ (psi), while Ting named it J. Thus, it came to be known as J/ψ.

Tau Lepton (1975): SLAC

The tau lepton was discovered by Martin Lewis Perl (b. 1927) with his colleagues in the SLAC in 1975. The tau lepton is the heaviest of all the discovered lepton and *it is the only lepton that can decay into hadrons*. In 1995, American physicists Martin Lewis Perl (b. 1927) and Frederick Reines (1918-1998) shared the Nobel Prize in Physics, with Reines awarded his share of the prize in the experimental discovery of the neutrino.

Bottom Quark (1977): Fermilab

In 1973, Japanese physicists Makoto Kobayashi (b. 1944) and Toshihide Maskawa (b. 1940), predicted the existence of the bottom quark and top quark to explain the observed CP (Charge-Parity) violation in kaon decay. (The name "bottom" was introduced by Haim Harari in 1975, although there were efforts to name the bottom quark as "beauty.")

In 1977, the team of Leon M. Lederman from Fermilab discovered the bottom quark.

W and Z Bosons (1983): CERN

The Z and W bosons are the particles that mediate the weak force. Signals of W boson were seen from the series of experiments in January 1983 by Carlo Rubbia (b. 1934) and Simon van der Meer (1925-2011) of CERN. Later, the name of the experiments and team called UA1 (led by Rubbia) and UA2 (led by Peter Jenni) found the Z boson in May. Simon van der Meer (1925-2011) and Carlo Rubbia (b. 1934) won the 1984 Nobel Prize in Physics.

Top Quark (1995): Fermilab

In 1995, the top quark was discovered by the CDF and DØ (sometimes written as D0 or DZero) experiment at Fermilab.

Tau Neutrino (2000): Fermilab

In 2000, the tau neutrino was discovered at Fermilab by the experiment called DONUT (Direct Observation of NU-Tau), beating CERN's detector called NOMAD (Neutrino Oscillation Magnetic Detector), which was designed, in part, to reveal the neutrino.

Note

Physicists think that the Standard Model should be a compact what-you-see-is-what-you-get stand-alone theory that can explain everything that we can observe in our world. For example, physicists observed that the universe is expanding, while galaxies don't fly apart and so coined the terms dark energy and dark matter to explain the respective phenomena. Physicists said that the Standard Model does not incorporate the dark matter and dark energy into the theory. Obviously it does not. But for the sake of an argument, if the Standard Model was discovered before electricity, the mechanism of electricity still had to be discovered to say that the Standard Model can explain electricity. That is, just because we do not know about the dark matter or dark energy does not mean that the Standard Model cannot explain it. All it needs is just to discover or explain the dark matter and dark energy in the context of the Standard Model. Just because it is not yet discovered right now does not mean that it can be found somewhere in what the physicists now called New Physics. The term New Physics seems to be the repository of everything that the physicists can think of and what they can't solve.

Part 2
The Theories

Chapter 3
Charge Theory

It is amazing how physics had reached these incredible advances in science without the benefit of understanding what charge really is. From the time the phenomenon of electricity was observed to the discovery of the charges of the subatomic particles (the positive charge proton, the negative charge electron, and the neutral charge neutron), charge is the defining property of our world. Somehow, in the midst of all these gathering of knowledge, the prevailing understanding of charge operates so that while it is not truly understood, it does not hinder the march of scientific discoveries. Thus this perception that there is nothing wrong with the understanding of charge eludes the immediate attention of the scientists. In order to understand charge, we have to go back to that early history of electricity, explain what charge is through a theory, apply this new theory to explain a phenomenon, and provide concrete proof which could be used to explain or solve any related contemporary scientific problems.

Thales of Miletus: Amber and Electron

Thales (640-546 BC), an Ionian Greek philosopher from Miletus in the Asia Minor (present day Turkey) was known as the first person to have studied electricity. Thales observed that when amber (which came from a fossilized tree resin) is rubbed with a woolen cloth, the amber would attract small, light materials such as lint, small pieces of straw, feathers, and bits of dry leaves. Due to this strange quality of amber, Thales believed that it contained some unknown force. The Greek word for amber is *elektron*.[1]

Benjamin Franklin: Naming of the Charges

In 1733, Charles du Fay (1698-1739), a French scientist published his discovery that there are two kinds of "charge," which was then called "electrical fluid." One that is produced by glass is called "vitreous," while the other that is produced by amber is called "resinous." Du Fay believed that these fluids can be separated by the friction of rubbing.

Fifteen years later, Benjamin Franklin (1706-1790), an American scientist and politician proposed that "vitreous" and "resinous" were not actually the different kind of electrical fluids, rather, in the process of rubbing, a single fluid flows from one body to the other. He coined the term "positive" for "vitreous" and "negative" for "resinous." That is, he designated the body that gained the fluid as *positively charged*, while the one that lost the fluid as *negatively charged*.

Franklin from his experiments coined many other electrical terms that we are still using today such as battery, charge, condensor, conductor, plus, minus, positively, negatively, and armature.

The Difference between "Charge" and "Charged"

As early as the start of this book I had differentiated between the term "charge" and "charged." "Charge" is an inherent property of a particle, be it positive, negative, or neutral—not to be confused with charge to effect the change of charge of an atom or an object. "Charged" is related to the *charge* of an otherwise mostly electrically neutral atom (or object), where it could be made as positively charged or negatively charged depending on its shedding or acquiring of electrons.

Experiments on Charges

One of the observable phenomena of charge is the visible attraction and repulsion of lighter objects. This can be demonstrated through an experiment using two Styrofoam balls suspended with a string and using a glass rod and a silk (used instead of amber) that will be used to charge the Styrofoam balls (Figure 3.1).

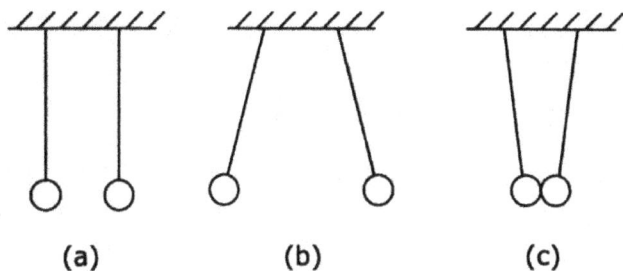

      (a)   (b)   (c)

Figure 3.1. The phenomenon of attraction and repulsion using a glass rod, silk, and two suspended Styrofoam balls. a) Styrofoam balls untouched by the glass rod or silk. b) When the two Styrofoam balls are rubbed by the glass rod (or silk), both Styrofoam balls repulse each other. c) When one Styrofoam ball is rubbed with the glass rod and the other with the silk, the Styrofoam balls are attracted to each other.

The attraction or repulsion is caused by the transfer of the electrons from one body to another. (Note: Glass–vitreous–positive/amber or silk-resinous-negative.) When the glass rod is rubbed on the Styrofoam, the electrons from the Styrofoam are transferred to the glass rod, making the Styrofoam ball positively charged. When silk is rubbed on the Styrofoam ball, the electrons from the silk are transferred on the Styrofoam ball, making the Styrofoam ball negatively charged.

So, what is charge? Quite simply, the charge: positive and negative, is just a convention much like what Benjamin Franklin named them. (The real understanding of charge will have to wait further down in the section on *Charge Theory*.) In a way it is a convention based from the observed property of an object as the basis of its charge in relation to other objects. That is, the amber, which was labeled as resinous, was labeled as negatively charged and any object that is repulsed to it is negatively charged, while any object that is attracted to it is positively charged. Charge, in this sense, is the property of an object to shed or attract electrons.

Charge (In Physics)

Another understanding of charge in physics can be seen in the definition of its unit. The SI unit of charge is coulomb, C, expressed in meter-kilogram per second unit of electrical charge, which is equal to the quantity of charge transferred in one second by a steady current of one ampere.

From the experiments of the scientists, it was observed that charge appeared only in discrete amounts, that is, it is quantized. In 1909, Robert A. Millikan (1868-1953), an American physicist first measured the quantum of charge approximately as:

$$e = 1.602 \times 10^{-19} \text{ C}$$

The charge (denoted by the letter q) must be an integer of this basic unit, e. That is, $q=0$, $\pm e$, $\pm 2e$, $\pm 3e$, etc. The magnitude of charge of electron and proton are the same:

$$q_e = -e \qquad q_p = +e$$

The understanding in physics is that the electron itself is not the charge, that is, charge is a property, like mass of elementary particles such as the mass of the electron.[2]

(In contradiction to the stipulation that the charge must be an integer value, in Chapter 2 on *The Standard Model*, the quarks contained within the proton and neutron carries a fractional charge of ±1/3 or ±2/3 of *e*.)

Spin (In Physics)

As discussed in Chapter 1 on *Electron Spin*, the spin of a particle in Quantum Mechanics is only a "quantum-mechanical phenomenon" and it has no counterpart in Classical Mechanics. That is, the spin of a particle is not the same as the planet spinning on its axis.

Charge Theory

The understanding of charge is one of the cornerstones of physics. Without it, physics cannot move forward.

My first understanding of charge was my observation of the interaction of the direction of the spin of objects. For example, two objects spinning (like two spin top, two gears, or two whirlpools) in a clockwise or counterclockwise would counteract (repeal) each other; while if one spins in a clockwise direction and the other in counterclockwise direction, they will aid (attract) each other. From these, it occurred to me that the spin (the direction of the spin or non-spin) of the particle is its charge. **That is, charge and spin are one and the same**.

To understand charge, I went first from understanding the spin of electron using the right-hand rule. The right-hand rule is used to determine the direction of the magnetic field around a wire conductor knowing the direction of the electrical current. This is done by using the right hand where the thumb points to the direction of the electrical current and the fingers wrap around the wire conductor shows the direction of the magnetic field. Say, in a wire conductor that is positioned straight up and the direction of the electric current is going "up" the wire, applying the right-hand rule shows that the magnetic field is in a counterclockwise direction looking down at the wire conductor. From the right-hand rule, the electron *seems* to spin in a counterclockwise direction.

Later, after seeing the bubble chamber images, I found that the right-hand rule of the electron does not agree with the spin of the charge particles. The bubble chamber images showed that negative charge particles spins in the clockwise direction, while that of the positive charge particle spins in the counterclockwise direction. (The neutral charge particle does not show up in the bubble chamber but its path is straight.) Which one is right? The right-hand rule with the electron (that seems to be) spinning in counterclockwise or the bubble chamber images where a negative charge particle showed as spinning in a clockwise direction. I scrambled to explain this discrepancy. Luckily, it did not took me long to find the answer. To gather all my thoughts and to analyze the problem, I went back to the understanding of the *Right-Hand Rule*.

The Right-Hand Rule and the Left-Hand Rule

The Right-Hand Rule (Direction of the Flow of Electric Current by Convention) and the Left-Hand Rule (Actual Flow of Electrons)

As explained, the right-hand rule is used to find the direction of the magnetic field around the wire conductor knowing the direction of the electric current. Luckily, I recalled that the direction of the electric current was just a convention. (It was agreed upon at the time when the actual flow of electron was not yet known. Also, it was done for computation purposes.) That is, the actual direction of the flow of electron is actually opposite to that of the "flow" of electric current.

As the direction of the magnetic field is the same in the right-hand rule, the only way to reflect the direction of the actual flow of electron is by using the left hand. This is called the left-hand rule. In the left-hand rule, with the same wire conductor straight up, the electric current is going up while the electron flow is going down. With the same direction of the magnetic field, the left thumb is pointing down following the direction of the electron.

Now, turning the left hand up agrees with the direction of the negative charge particle in the results of the bubble chamber. That is, negative charge particle spins in the clockwise direction. Reflecting this on the positive charge particle with the right-hand pointing up has the positive charge particle spinning in the counterclockwise direction.

So to summarize: Negative charge particles follow the left-hand rule, while positive charge particles use the right-hand rule.

From the reconciled spin of the electron in the wire conductor and the results of the cosmic rays in the bubble chamber, I am proposing the Charge Theory as follows (Figure 3.2):

Postulate #1: Negative charge particle spins clockwise.
Postulate #2: Positive charge particle spins counterclockwise.
Postulate #3: Neutral particle do not spin.

This simple theory on charge can now be compared with the results of the existing experiments on the bubble chambers and cloud chambers and the theory will be found to agree them.

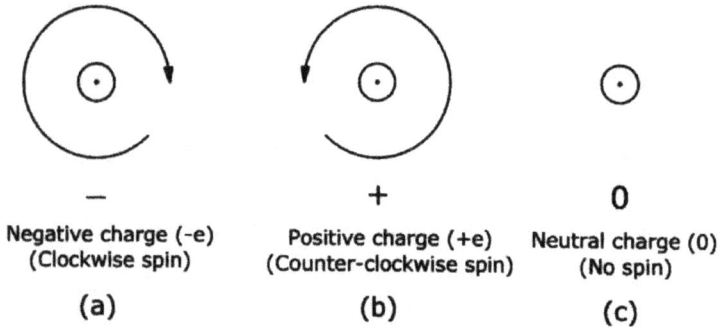

Figure 3.2. Charge Theory. Particles viewed from above. a) Negative charge b) Positive charge c) Neutral charge.

Charge—Defined

The term charge can now be defined with the following understandings:

- Charge is the direction of the spin of particle for the charge particles and non-spin of the neutral charge particle.
- Charge is the attraction of two unlike charge particles or the repulsion of two like charge particles.
- In terms of electrons or electric current: Charge is the flow of electrons from the material with more electrons to the material with fewer electrons.
- Charge is the direction of the spin of the field.
- Charge is the (value or) strength of the spin of the field.

Charge in its outward manifestation is about repulsion and attraction. In matter, it is more about the flow of the negative charge electrons as the electrons are the only free particle that could move in an atom.

In Chapter 5 on *Quark Theory*, I came to understand that the whole charge value of the electron (-e) and the fractional charges of the quarks (-1e/3 for down quark and -2e/3 for up quark) refers to the *strength of the spin of the field*. That is, the field of the particle such as the electron and the quarks spins just like the particle itself. In the proton and neutron, the field of the proton spins, while that of the neutron does not spin.

Application of Charge Theory on the *Experiment on Charges*

The positive and negative charges are differentiated by the direction of the spin of the particle, that is, the negative charge electron spins clockwise, while the positive charge proton spins counterclockwise. Particles of the same direction of spin repel each other, while particles of opposite spin attract each other. This can be applied on the previous section on the *Experiment on Charges* (see Figure 3.1), where the Styrofoam rubbed with glass rod (vitreous/positive) became negatively charged, while the other Styrofoam rubbed with silk (resinous/negative) became positively charged (Figures 3.3, 3.4, and 3.5). Silk or amber is resinous or negatively charged since it can contribute more electrons to the object it is in contact with, while the glass rod is vitreous or positively charged since it can accept or grab more electrons from the object it is in contact with. Note that the negatively charged object will have more electrons, while that of the positively charged object will have less electron, thus will have more protons or is positively charged.

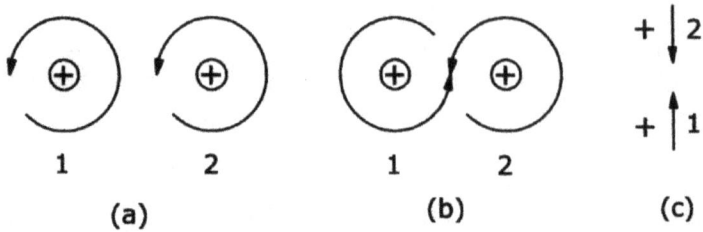

Figure 3.3. From Figure 3.1b: Silk touched to the Styrofoam balls. The electron from the silk moved to the Styrofoam balls making both balls negatively charged. a) Two negative charge particle spins in a clockwise direction. b) Rearranged drawing of the spin of particles. c) Both spin of particles represented by an arrow that are opposing each other.

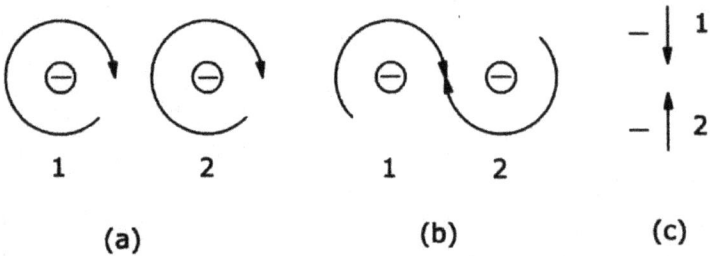

Figure 3.4. From Figure 3.1b: Glass rod touched to the Styrofoam balls. The electrons from the Styrofoam balls moved to the glass rod making the Styrofoam balls positively charged. a) Two positive charge particles spin in a clockwise direction. b) Rearranged drawing of the spin of the particles. c) Both spin of particles represented by an arrow that are opposing each other.

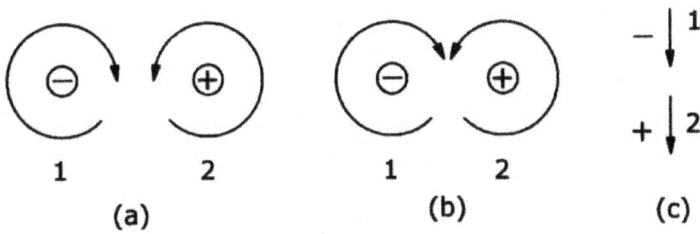

Figure 3.5. From Figure 3.1c: One Styrofoam ball is touched with the silk and the other with the glass rod. The Styrofoam ball touched by the silk becomes negatively charged while the one touched with the glass rod becomes positively charged. a) A negative charge particle in clockwise direction, while the positive charge particle in a counterclockwise and clockwise direction. b) Rearranged drawing of the spin of the particles. c) Both spin of particles represented by the arrow that are in one direction or aiding each other.

Application of Charge Theory on Cathode Ray Tube with an Electric Field

From the experiment on cathode ray tube, the straight streams of the cathode rays (the flow of free electrons) can be bent by the negative side of the electric field where the electrons travel freely from the negative terminal to the positive terminal (see Figure 1.2). This observation can be explained by the Charge Theory simply with the spin of the negative charge particles, the electrons (Figure 3.6).

What Figure 3.6 shows is that the higher energy electrons of the electric field repulse the cathode rays. Also, it shows that even at a right angle direction of the source of the flow of the free electrons, they will still repulse each other.

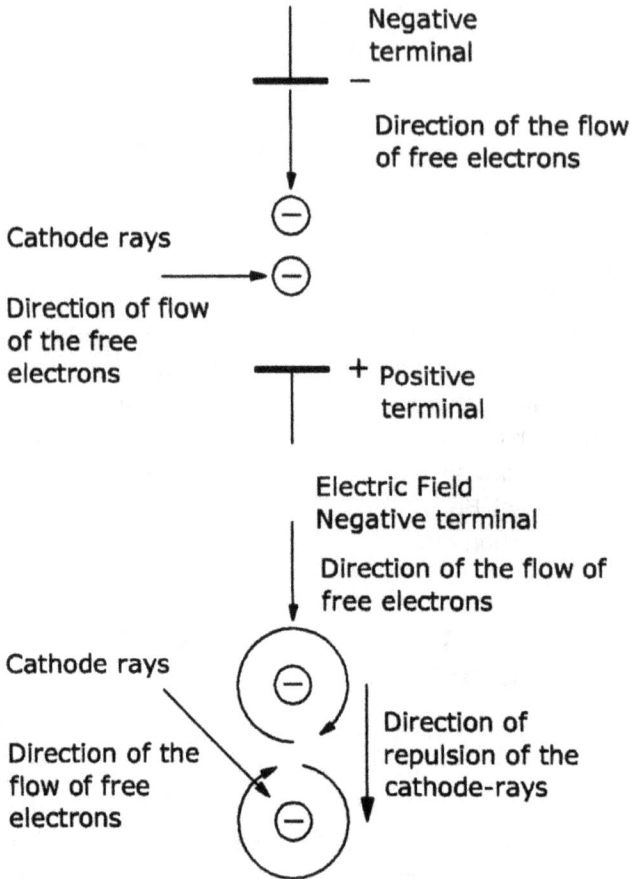

Figure 3.6. The bending of the cathode rays by the electric field. Top figure: The electrons from the negative terminal of the electric field flows towards the positive terminal of the electric field. Bottom figure: Both electrons from the cathode rays and the electric field repulse each other with the electrons from the electric field pushing the electrons towards the positive terminal, bending the path of the cathode rays.

Charge of Particles in the Bubble Chamber Images

The theory that charge is the same as the direction of the spin of the particle may only stay in the realm of a hypothesis in the explanation of the attraction or repulsion of the charge particles in the *Experiment on Charges* and in the repulsion of the cathode ray tube of the electric field without the benefit of a visible proof. For this reason, observation of a phenomenon and a theory that explains the phenomenon could gain a much better support for the theory if one could find a proof that is starkly visible. The most graphical proof that reinforces my Charge Theory is an example from the bubble chamber image (Figure 3.7).

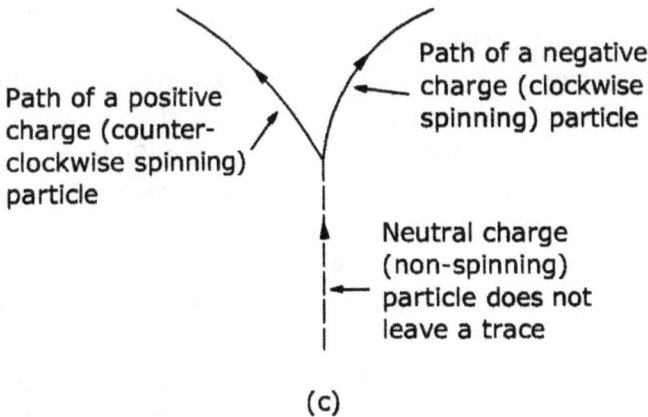

Path of a positive charge (counter-clockwise spinning) particle

Path of a negative charge (clockwise spinning) particle

Neutral charge (non-spinning) particle does not leave a trace

(c)

Figure 3.7. A typical bubble chamber image that shows a particle (photon with the energy of a gamma ray) producing a positive charge particle (positron) and a negative charge particle (electron). (The actual bubble chamber image is said to be an iconic image as it showed the direct conversion of energy into matter.)

(To be strict with the understanding of $E=mc^2$, everything is actually energy. There is no "conversion" from one form to another; there is only the result of reaction. Matter is just the manifestation of an energy particle that is moving and is

79

radiating its field. I am still questioning the idea of the observed result that an energy particle such as the photon could become the positron and electron. Rather, I would like to think that the positron and electron were just the product of the photon collision with the atoms of the bubble chamber. Also, since the positron does not occur normally in nature but is the product of the collision, I believed that the positron could be an electron that was forced to somehow been dislodged from the atom with such a force that it spins in the opposite direction.)

Physics books often draw a reproduction of the bubble chamber photograph with the path left by the particles labeled. The identification of the charge of the particles is very much consistent with the results of the bubble chamber images.

Just what Figure 3.7 shows, it is necessary to determine the direction of where the source particle came from to determine the charges of the resulting particles since this is the only way to determine the direction of the spin of the particle. (Without knowing the polarity of the magnetic field of the bubble chamber used to produce the images and with the observation of the consistency of the identification of the charge particles from the various bubble chamber images, it can be fairly said that the set-up of the magnetic field of the bubble chambers are standard.)

Some of the particles in the bubble chamber will trace a very big curve but still their charges can be determined by the direction of their curve. Some of the particles of the bubble chamber will also trace a curl getting smaller (Figure 3.8). A positive charge particle such as a positron (Figure 3.8a) is said to be "curling to the left," while a negative charge particle such as an electron (Figure 3.8b) is said to be "curling to the right." What the description of a particle that says "curling" *in a certain direction* does not give the right understanding of what charge is, except only as a description of the direction of the motion of the particle.

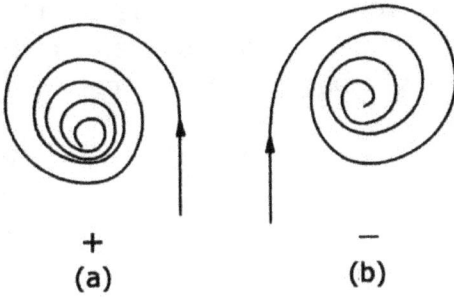

Figure 3.8. Charge particles in the bubble chamber images. (a) A positive charge particle "curling to the left." (b) A negative charge particle "curling to the right."

What Figure 3.8 also shows is that a high-energy particle traces a much bigger curve path but as the particle loses energy the trace of its curve or its curl grows smaller. (A much better understanding than the argument that the energy of the particle remains the same while that its spin slows down.)

Summary

The early understanding of charge was the ancient Greek's discovery of the property of amber to attract small light objects (the repulsive and attractive property of an object) and to Benjamin Franklin's understanding that charge were electrical flow from one object to another. Charge was later understood to be the transfer or flow of the electrons from one object to another that could make other object negatively charged (more electrons) or positively charged (less electrons or more protons compared to the electrons). That is, charge is understood according to its observed phenomenon. Charge was defined also as the rate of flow of electrons and was given a value. That is, charge in physics was abstracted from its true understanding and was never understood to be the spin of a particle. The ultimate proof that the charge of a particle is its direction of spin is in the images of the result of the bubble chamber. Nothing more could be more compelling than that as the true understanding of charge was all along right before our eyes— hiding in our plain sight.

It is to be pointed out that even before I had understood the implication of the images of the bubble chamber, I had already associated the charge of the particle as its (direction of) spin from my aforementioned discussion in the section on *Charge Theory*—the bubble chamber images just reinforced what I had already known.

Chapter 4
New Model of an Atom

The New Model of an Atom is a model of an atom that incorporates the understanding of charge (Charge Theory), the mass of the quarks and subatomic particles (Mass Theory), and the Theory on the Structure and Mechanism of an Atom. The New Model of an Atom goes far beyond Bohr's Model of an Atom (which was based only on the hydrogen atom) by showing the structure of the nucleus from the hydrogen atom to isotopes of hydrogen atom (deuterium and tritium) to helium that is a stepping stone to the possible understanding of the structure of the atom of the rest of the elements.

* * *

We have come a long way from the acceptance of the reality of the atom with the help of Einstein's explanation of the Brownian Motion to the fix orbit of an electron of the Bohr's Model of an Atom to the probabilistic orbit of an electron of the Schrödinger's Model of an Atom. It is significant to point out that the passing from Bohr's Model of an Atom was marked with the changing from the term "orbit" of electron to Schrödinger's Model of an Atom using the terms "electron cloud" and "orbitals." Pauli Exclusion Principle guided the explanation of the Periodic Table of Elements with its arrangement of the electrons in the shell and subshell of an atom of an element. Not easily noticed is that almost all physics books often left out the Schrödinger Model of an Atom. It is quite strange that while the model of an atom described in shell and subshell and probabilistic movement of the electrons in an atom, Bohr's Model of an Atom was not totally discarded. Yet, all seems to be working fine with the structure of the atom.

With the development in physics of finding the most basic particles of matter, physics passes from the atom and its subatomic particles (proton, neutron, and electron) to the theory called Standard Model that pertains to the fundamental particles and forces in nature. So far, what had pre-occupied physics for the longest time (more than 40 years now) is the question of where do the particles got their mass as the theory Standard Model does not take it into account.

Unwittingly, these undeclared, unrecognized, and misunderstood problems of science are coming to the fore at this time. The answer to these problems provides sweeping explanations that will give light to science as it passes through its new age. What we will find is that the search for what gives mass to a particle is the understanding of the *mechanism* of how a particle gets its mass rather than finding another particle that gives mass to another. More so, the understanding of how does matter gets its mass is the understanding of the structure of the atom and the mechanism by which the subatomic having a mass of their own attains a mass as a whole as an atom of an element. With it, the true solution should be able to mesh with our existing correct theories and knowledge, reinforcing our understanding, and making the world much clearer. In order for all these things to pass, we have to go back to the very beginning—to a basic understanding of science.

Basic Physics

My first subject on physics was in my fourth year in high school where I barely remembered anything except what our teacher (which I had mentioned in the *Acknowledgements*) had taught us about how to understand an equation, such as the equation of *speed*.

Speed is the relationship between the *time* to cover a *distance* as shown by the equation:

v = d/t

where *v* is the speed, *d* is the distance, and *t* is the time.

In understanding this equation, we can say that: For the same distance, if we increase the time, the speed decreases; likewise, if we decrease the time, the speed increases. This is the same method of understanding and analyzing any equation that I learned later in engineering.

Later, when I wrote my first book on physics, the same understanding of the equation had guided me in trying to understand one of the greatest and most famous equation from Albert Einstein, the formula:

$E=mc^2$

where *E* is energy, *m* is the mass, and *c* is the speed of light.

The formula could be rewritten as $m=E/c^2$.

Applying the method of understanding an equation gives us the following: With the energy *E* remaining the same, if we increase the speed *c*, then the mass *m* decreases; likewise, with the energy *E* remaining the same, if we decrease the speed *c*, then the mass *m* increases. Applying this idea to a moving energy particle, what this means is that the mass of the particle came from its energy and motion. (This is obviously different from the usual prevailing understanding of $E=mc^2$ that matter and energy is interchangeable, and that a small amount of matter could produce an enormous amount of energy.) **That is, the mass of a particle is not *given* to it by another particle; rather it is an inherent property of an energy particle that is in motion.**

(As it is, I had been reusing the things that I wrote in my first book, my theory on light; as my journey towards the understanding of where the mass of the particles came from the understanding of the nature of light.)

Idea from My Unpublished Theory on Light

From the idea on my theory on light with regards to matter, I had wondered how the light, c, in the $E=mc^2$ was "captured" inside the atom. (It was the *way* I *understood* it before, the c as light instead of the speed of light.) I thought earlier that light acts as a "glue" that creates a "drag" to decrease the speed of the subatomic particles in the atom. This thinking would later lead me to my earlier understanding on what gives mass to an *atom*.

The Mass of Photon (Light) and Neutrino

It is at times ironic that physicists are still arguing if light has a mass while not considering that light exerts a force called radiation pressure where the photon of light attains a momentum. (Radiation pressure is the ability of light to push an object. One of its applications is on an outer space vehicle with a "solar sail" that propels the vehicle in space.)

When the physicists were asked if light has a mass, they would answer that the rest mass of the photon is zero. The right answer of course is that light has practically no *rest mass* since it only exists as a speeding energy particle. It does have a *mass* as it is an energy traveling with a speed—$m=E/c^2$. (The formula $E=mc^2$ actually refers to a particle moving in a vacuum.) The same understanding could be derived from the neutrino, which is taking the attention of the physicists right now.

Mass Theory: Theory on the Source of Mass of the Particles

The mass of a particle is its inherent property derived from its motion or speed. (We can also argue that **all** particles cannot exist without moving.) These are the ways that the particles had attained their mass:

1. Free particles or particles travelling in free space. This pertains to photon and neutrino. The difference between a photon and neutrino is that a photon spins (charge), while neutrino does not spin (neutral).
2. Particles spinning on their axis. This pertains to the quarks, proton, and electron. Electron is unique in that it spins and also orbits the proton.
3. Particles orbiting a nucleus. This pertains to the electron and to the down quark inside the proton and neutron (discussed in Chapter 5 on *Quark Theory*). Both particles are spinning on their axis and orbiting a center.

The question that should be asked at this point is: "Why is neutron that is supposed to be not spinning has a mass?" This will be answered later in Chapter 5 *Quark Theory*.

Where did the Physicists on the Standard Model Made the Mistake in the Understanding of How the Particles Got Their Mass?

In most discussions about how the particles got their mass (and why a theory, the Higgs Mechanism, was invented to provide the answer to this problem) was that the Standard Model does not account for the mass of the particles. The mediating particles of the fundamental forces of the Standard Model such as the photon, the gluons, and the W and Z bosons are supposed to have no mass. (Subsequently, the neutrinos are supposed to have also no mass to the extent that the mass of the whole fundamental particles are also explained by the theory of the Higgs Mechanism.) The problem occurred when the W and Z bosons were found to have very heavy masses, 80.4 GeV/c^2 and 91.2 GeV/c^2, respectively—which are about 100 times the mass of proton. Physicists scrambled to find an answer to explain this seeming disagreement.

(The Standard Model was the result of the understanding of the particles and forces, and these forces are mediated by the particles called bosons. As pertaining to the known particles and forces, these were observed from the experiments and so there is no question as to the reality of their existence. *In the field of particle physics where a particle was observed from the collisions of particles, to some extent, there is really no argument as to the existence of the particle—only as to its identity.*)

So where did the physicists made the mistake on the Standard Model in its understanding of how the particles got their mass? As I reached this point of understanding in time, I traced the misunderstanding to the question of, "Does a photon have a rest mass?"—which the physicists really don't know answer, of which the answer is none. But the photon, as an energy particle, does have a mass derived from its speed. Then it occurred to me that the proton, neutron, electron, as all the quarks derived their masses from spinning on their axis or orbiting a nucleus, that is, they are all practically moving in place. The mediating forces such as the photon on the other hand move between the electron and proton (the nucleus), and photon as a free particle move in space attained their masses through their inherent existence as a speeding particles, hence the bosons are understandable to have no rest mass.

So why are the W and Z bosons very heavy? The answer to this maybe is that the W and Z bosons are the carrier or mediating force between the nucleus and the electron—or that the W and Z bosons are the particles that are holding the electron to the nucleus. The masses of the W and Z could have been derived from the amount of energy needed to "decouple" the forces that holds the electron to the nucleus together.

The Evolution of My Idea on the Structure of an Atom

Bohr's Model of an Atom was founded on the idea that the electrostatic force between the nucleus and the electron was what causes the centripetal force that made the electron orbit around the nucleus. Since the physical spin of the particles are not recognized but rather regarded as just as a quantum number, the electron and proton was never depicted to spin even in the depiction of an atom. How I had reached the knowledge of the structure of the atom was the result of my series of discoveries and the evolution of my idea on the structure of an atom.

The Dipole Particle

The idea that a subatomic particle is a dipole particle, that is, it has a "North pole" and a "South pole," had been around for quite some time (Figure 4.1).

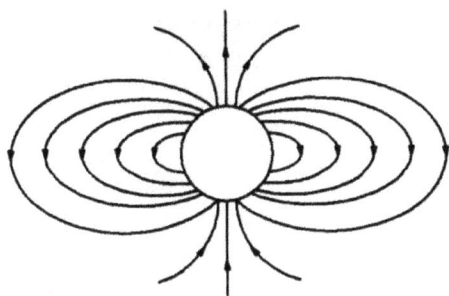

Figure 4.1. Dipole particle.

In my theory on the structure of an atom, the subatomic particles (proton, neutron, and electron) and the quarks inside the proton and neutron are dipole particles and their spin is a real mechanical motion of the particle.

My Idea that the Proton is the One that Forces the Electron to Orbit Around It

In my theory, there is only a one to one assignment of an electron to the proton and that the proton is the one that makes the electron orbit around it. Thus, unlike the Sun where many planets orbit around it or some planets where many moons (or satellites) orbit around it, there is one and only one electron that can orbit a proton.

My Early Idea of the Structure and Mechanism that Created the Mass of the Atom

My early idea of the structure of an atom was not based on my search of the structure of an atom, rather it was based on the source of mass of an atom based on my understanding of $E=mc^2$. In my earlier understanding of $E=mc^2$, I thought of light as a "glue" in matter; I thought of light as something that creates a "drag" inside the atom to create a mass. That is, I was looking for the arrangement of the subatomic particles in the atom that which produces the "drag." Based on the current knowledge of an atom, the proton and neutron have no arrangement within the nucleus. As I applied the idea of something that acts as a "drag" to the atom, I placed the non-spinning neutron around the proton (Figure 4.2).

In the scheme of Figure 4.2, I envisioned the neutron that is not spinning on its axis in two ways: One is that the neutron is orbiting the proton, and the other is that it is fixed in its place. My thought is that the neutron creates a drag on the spin of the proton. This was my idea for quite some time until another idea corrected it. (This will be discussed further down in the section *New Model of an Atom*.)

90

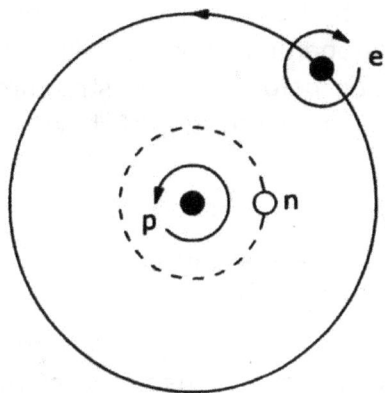

Figure 4.2. My early idea of a mechanism that produces the mass of the atom.

My Idea of the Nuclear Fusion from the Sun

My idea of the nuclear fusion of the hydrogen atoms to produce the successive isotopes of the hydrogen (deuterium to tritium) until it produces helium is that I visualized the hydrogen atom being stack on top of each other with the hydrogen atom shape like a pancake. In this process, the immense pressure squeezes the two hydrogen atoms where I imagine the electron of the hydrogen atom *below* the two stacked hydrogen atoms is squeezed out, rendering the proton to a non-spinning neutron. (This will be discussed more in the succeeding section on *New Model of an Atom* below.)

Creation of the Elements from Hydrogen to Helium

Having stumbled earlier upon the possible solution to the structure of an atom, I search at the beginning of how the atom of an element was created. Naturally, the elements I studied were hydrogen and helium atoms. Looking at the depiction of the hydrogen and helium atom, it occurred to me that there was a "jump" on the creation from hydrogen to helium atom. I went to look at the isotopes of hydrogen that I am familiar with but had not paid any close attention before on their structure. The following is my journey to my discovery of how the elements were created.

Current Depiction of the Atom of the Elements: Hydrogen, Deuterium, Tritium, and Helium

In order to have an idea in the steps of fusion of the hydrogen atoms of the Sun, it helps to see the current depiction of the hydrogen atom to its isotopes to the helium atom (Figure 4.3).

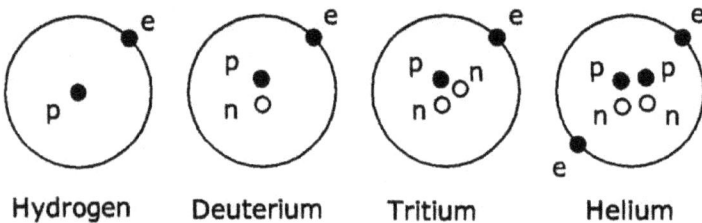

Hydrogen   Deuterium   Tritium   Helium

Figure 4.3. Fusion of the hydrogen atom forming the successive isotopes of deuterium, tritium (which is radioactive), until finally the helium atom.

Nuclear Fusion in the Sun of Hydrogen into Helium

The explanation of the nuclear fusion in the Sun is usually explained as the fusion of the hydrogen atoms inside the Sun to produce the helium atom. To understand the process or mechanism of the nuclear fusion of the hydrogen atom to produce helium, the best way to do this is to study the hydrogen atom and its next isotope, the deuterium atom (Figure 4.4).

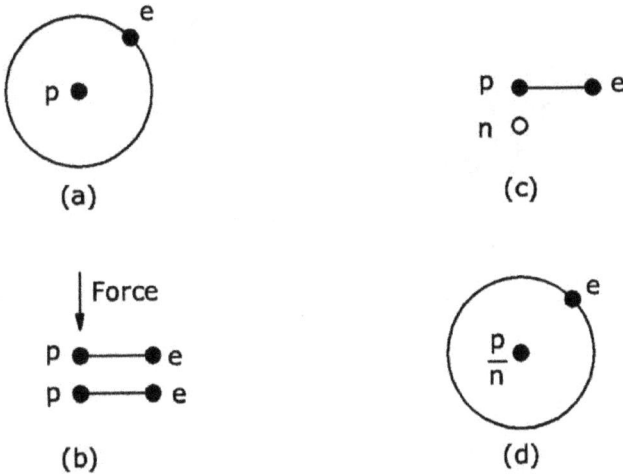

Figure 4.4. Nuclear fusion of two hydrogen atoms into deuterium. a) The top view of hydrogen atom. b) Side view: Immense pressure compresses two hydrogen atoms. c) Side view: One hydrogen atom loses its electron, while its proton "turns" into neutron, releasing an amount of energy. d) Top view of (c): The resulting atom of two fused hydrogen atom after the fusion reaction.

In this nuclear fusion, one hydrogen atom was "squeezed" to "release" the electron and turn its proton into neutron releasing an electron and an electron antineutrino.

Structure of the Atom of the Elements: Hydrogen, Deuterium, Tritium, and Helium

Based on the idea derived from the discussions above and the figure shown by Figure 4.4, this could now be applied on the hydrogen, deuterium, tritium, and helium shown in Figure 4.3 to derive how the structure of their atom look like (Figure 4.5).

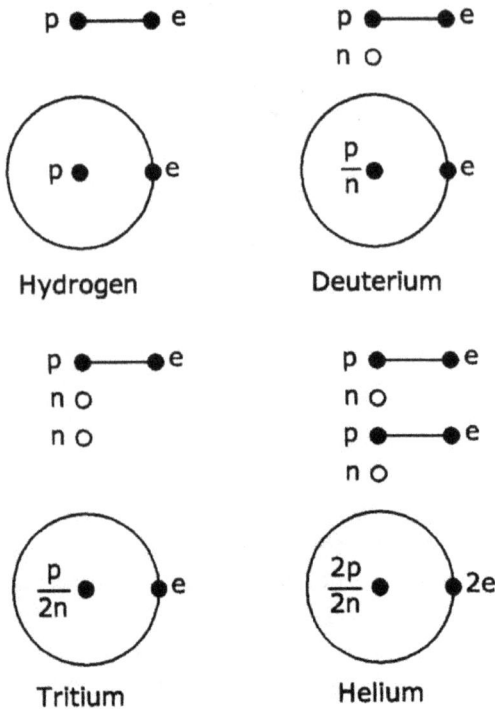

Figure 4.5. Structure of hydrogen, deuterium, tritium, and helium atoms.

The idea that can be derived from this pattern is that the hydrogen atom seems to lose the electron alternately from the succeeding nuclear fusion. This pattern seems to hold true around the element calcium where after that the number of neutrons becomes greater than the number of protons. Also, the nucleus of an atom is not lump as the current understanding of the model of an atom but rather is structured as pole of protons and neutrons.

New Model of an Atom: Charge Theory, Mass Theory, and the Theory on the Structure and Mechanism of an Atom

The New Model of an Atom shows the charge of the proton (spinning counterclockwise) and the electron (spinning clockwise), the electron being made to orbit by the proton in the direction of the spin of the proton (counterclockwise), and the structure of the nucleus consisting of proton and neutron in a proton-neutron pole (Figure 4.6).

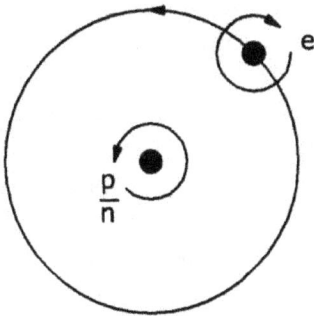

Figure 4.6. The New Model of an Atom: The positive charge proton or counterclockwise spin of proton forming a proton and neutron pole, negative charge electron, and direction of orbit of electron.

This New Model of An Atom shows that the proton and the neutron are arranged into what I called the proton-neutron pole and are not lump together like the understanding of the Schrödinger Model of An Atom and even the Nuclear Shell Model of an Atom. The neutrons by virtue of the quarks inside it, much like the proton should have the same "North" and "South" poles as the proton. Without even taking into account the presence of the neutron, the protons are still attracted to each other through their opposite poles much like the bar magnets could be made into a pole by attaching their opposite poles. In this sense, there is really no repulsion between the protons in an atom, and that the strong nuclear force, which was supposed to prevent the repulsion of the protons with each other, is actually the force that connects the protons through their opposing poles. This mediating or carrier force of the strong nuclear force is currently recognized as the pion—under the hadron called meson.

Why Physicists Cannot Come Up With the Solution to the Structure of an Atom

The following are the reasons why the physicists cannot come up with the solution to the structure of an atom:

1. The electron around the nucleus turned from planetary-like orbit to the probabilistic location of the electron in the region called electron cloud. This was caused by Schrödinger's Model of an Atom.
2. The proton and neutron of an atom is lump together as a nucleus with no order. The protons are supposed to repulse each other but are prevented from flying apart by the strong nuclear force.
3. Physicists did not recognize from the Periodic Table of the Elements that there is a one to one relationship between the proton and electron.
4. Physics do not understand that the charge and spin are one and the same.

Physicists cannot easily overcome these limitations since they were bound by these accepted ideas and misunderstandings that made up the contemporary knowledge.

On Schrödinger Equation and Heisenberg Uncertainty Principle

Schrödinger had been influential in changing the early structure of an atom from an electron orbiting the nucleus much like the solar system, to the so-called electron cloud where it became a matter of probability to locate the electron. (Bear in mind that Schrödinger believed that a particle is really a group of waves, a wave packet, somewhat like a fuzzy powder puff, which had given rise to the idea of an electron cloud.) Schrödinger and Heisenberg had been influential in introducing probability in Quantum Mechanics.

My New Model of an Atom practically demolishes the idea of the Schrödinger's Model of an Atom in the sense that the electron was returned to the idea that it is in orbit around the nucleus, or rather more exactly, it is in a fixed orbit around a proton.

As with regards to the possibility of finding an electron that orbits the nucleus, it is technically possible to "catch" the electron due to the speed of the orbit of the electron as anywhere we put our detector, there is where the electron is. That is, we are limited by the physical dimension of the atom and the speed of the electron in detecting its position.

In a sense, Heisenberg's Uncertainty Principle follows that of Schrödinger's Equation by positing the impossibility that we can practically determine the location of the electron of an atom. The bottom line of these "equation" and "principle" is that they state the commonsense truth or the obvious fact (as all laws of nature are—after discovering them), yet they abstracted the understanding that the electron in an atom that we are observing requires us to use materials made of matter, that is, materials made of atoms that practically affects the state of the electron in an atom that in turn affects the state of the material "detector." That is, there is no way we can passively observe an electron without disturbing it.

Summary

In this chapter, I had shown that the solution as to how the particles got their mass had been with us all along in the formula $E=mc^2$, where mass is an inherent property of a moving particle whether by spinning, travelling, or moving in orbit.

In the process of the fusion of hydrogen to helium in the Sun, I had shown that the isotopes of hydrogen, which are the intermediate atoms between hydrogen and helium, showed us a glimpse of the structure of the atom of the elements.

In the New Model of an Atom, the atom shows the charge (spin) of the proton and neutron, the structure of the atom (arrangement of the proton, electron, and neutron), and the mechanism (the proton is responsible for the orbit of the electron). Unlike Bohr's Model of an Atom, which is based on the hydrogen atom of one proton and one electron, the New Model of an Atom specifically describes the location of the neutron in the nucleus.

Chapter 5
Quark Theory

Quarks are confined within the hadrons and never found in isolation, hence there is no known structure of the quarks inside the proton and neutron. Quarks are instead shown in various ways such as depicted in Chapter 2 on *Depiction of Quarks inside the Proton and Neutron* (refer to Figure 2.2). Is there a way to theorized on how the quarks are structured inside the proton and neutron? The Quark Theory discussed below could possibly challenge the prevailing theory on the quarks, the Quantum Chromodynamics (QCD).

Quark Theory: Theory on the Quark Structure of Proton and Neutron

Having the idea on the structure of an atom and charge, I used this knowledge and apply it to the idea of quarks inside the proton and neutron (Figure 5.1).

The charge of the quarks inside the proton (+2e/3, +2e/3, -1e/3) totals to +e, while the charge of the quarks inside the neutron (+2e/3, -1e/3, -1e/3) totals to zero (neutral).

In my Charge Theory, the spin of the particle is its charge. The question is, "Why do the quarks have fractional charge?" The ready answer could be to satisfy the arithmetic of the charge of the proton and neutron. My earlier thoughts was to make the spin an integer value as I thought that fractional value of spin does not make sense, and so I gave an alternate values to the proton (+3e); neutron (0); electron (-3e); up, charm, top: (+2e); and down, strange, bottom: (-e).

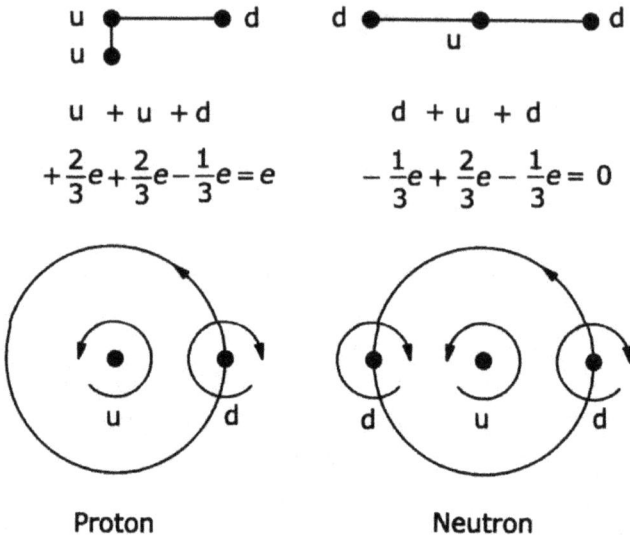

u + u + d

$$+\frac{2}{3}e + \frac{2}{3}e - \frac{1}{3}e = e$$

d + u + d

$$-\frac{1}{3}e + \frac{2}{3}e - \frac{1}{3}e = 0$$

Proton   Neutron

Figure 5.1. Structure of the quarks inside the proton and neutron.

There is another way to keep the values of the charge of the particles (quarks, electron, proton, and neutron) by the following understanding of the charge of the particle:

1. The direction of the spin of the particle.
2. The direction of the spin of the field.
3. The (value or) strength of the spin of the field.

Note that the proton and neutron are composed of quarks but the aforementioned understanding of charge that is primarily applied to the quarks and lepton could also be applied to the subatomic particles, neutron and proton.

In proton and neutron, their quarks always spin on their axis and that the *down* quark always orbits the *up* quark. The resultant fields of the *up* and *down* quarks formed the field of the proton and neutron. The field of the proton spins counterclockwise, while the field of neutron does not spin.

The understanding that charge is also the value or the strength of the spin of the field would answer why the arithmetic of the charges works for the proton and neutron so as to make the proton spin in the counterclockwise direction and the neutron to not spin.

Mass and Spin of Neutron: An Argument for the Existence of Quarks

When I was thinking of how the neutron got its mass, I thought I had hit the wall. (My theory of how the particles got their mass was discussed in Chapter 4 *New Model of an Atom*.) We can simply think that the proton got its mass since it is a spinning particle. The problem is that neutron, being neutral, should not be spinning at all on its axis and naturally based on my Mass Theory—it has no mass. I thought that it would take me a while to find the solution. Luckily, the solution came that reinforces the reality of the quarks. That is, referring to Figure 5.1, the *up* and two *down* quarks inside the neutron spins but since the spin of their fields neutralize each other, the neutron as a whole does not spin and hence has no charge. This neutralization of the resultant spin of the field of the quarks that produces the field of the neutron may have an effect on the spin of the quarks and could be the reason why neutron is much heavier than proton.

Arithmetic of the Charges

My first idea of the arithmetic of charges is about attraction and repulsion on my Charge Theory, which is shown in Figure 5.1. In the quarks of the proton, there is the attraction between the *up* and *down* quarks, which I arithmetically added without thinking too much about their charges. I did the same with the quarks of the proton and neutron. It took me a while to make sense of the charge. At this stage of my understanding, I had learned that:

- Charge is spin
- The (value or) strength of the spin of the field

The arithmetic of the charges applies to the charge of the atom and the charge of the proton and neutron. In the arithmetic of the charges, the attraction and repulsion changes to the idea of aiding and opposing, respectively. Applying this to the subatomic particles of the atom or the quarks inside the neutron and proton takes the following steps:

1. "Superimposing" of the axis of the particles into one
2. The arithmetic sum of the charges of the subatomic particles or the quarks is the resultant charge of the atom, neutron, and proton—which is their spin or the value of their charge

In the New Model of an Atom, which explains why there is an equal proton and electron in an atom in the Periodic Table of the Elements (which I will discuss in Chapter 6), shows that an atom is electrically neutral (charge) or as a whole its field is not spinning. In the proton and neutron, the field of the proton spins in counterclockwise direction, while that of the neutron does not spin—the same as the electrically neutral atom.

Theory on the Order of Quarks and Leptons

As I.I. Rabi once said upon hearing of the discovery of muon, "Who ordered that?" If he was still here right now, I would say, "Exactly!" Well, not really in the sense of what he meant. Rather, what I meant is in the term "order."

Our universe is practically just made up of the *up* and *down* quarks, which composed the proton and neutron, and the electron as the lepton. My theory is that these are the only fundamental particles: the up quark, down quark, electron, and the rest are the higher order or higher energy of these particles. (In the current Standard Model, the neutrino is classified as a fundamental particle, although I actually lean towards the idea that neutrinos are much like the photon, hence a boson—a free non-spinning energy particle.)

How did I come into this understanding? The following observations had contributed to this hypothesis:

1. In a purely unscientific way, in the early observation of the Standard Model, I had observed that the term *up* quark is similar to the *top* quark (up=top) as well as the *down* quark to the *bottom* quark (down=bottom). (Is it charmingly strange how the physicists had named them?)
2. The *up*, *charm*, and *top* quarks have the same value of +2/3 charge, while that of the *down*, *strange*, and *bottom* quarks have the value of –1/3 charge.
3. The increasing masses of the quarks from *up* to *top*, *down* to *bottom*, electron to tau, and electron neutrino to tau neutrino.
4. I had also observed that the succeeding heavier quarks were discovered with the increase in the energy of the machines.

One of the most important observation on this regarding the particle accelerators is that increasing their energies will hypothetically yield higher masses of the orders of the *up* and *down* quarks. For example, we can roughly graph the masses of the up, down, charm, strange, top, and bottom (refer to Table 2.1), and obviously, we are graphing their masses according to the energy of the particle accelerator needed to produce these particles (Figure 5.2).

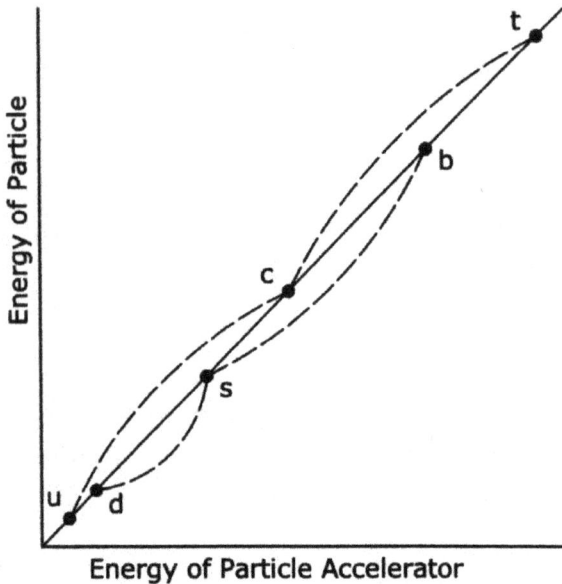

Figure 5.2. Graph of the masses of the quarks (order: u, c, and t; order d, s, and b) and the energy of the particle accelerator needed to produce them.

What the graph implies is that we can theoretically have quarks with a mass higher than the order of *top* and *bottom* quarks but they can only be achieved with higher energy particle accelerators—which entails higher cost of construction. It is just a question of how far are we going to go to prove it. I believe that our current particle accelerators should confirm this theory on the current existing discovered quarks so that this argument can be settled once and for all. (Fermilab's decommissioned Tevatron, if resurrected, has the same chance of proving this theory as particle accelerator Large Hadron Collider (LHC) of CERN (European Organization for Nuclear Research).)

Pion: A Meson or a Boson?

For quite some time I had recognized the confusing role of the pion, which is classified as a hadron under the meson, containing two quarks (a quark and an antiquark), while on the other hand, pion is also the mediating particle of the strong nuclear force within the nucleus and as such it is a boson.

So what is a pion really? Is it a meson or a boson or both? I believed that the pion is really a boson—which brings into question whether the pion actually have quarks inside it (as does all the mesons). That is, the pion can be just a pure energy particle much like the photon, electron, and the quarks.

Questions on the Baryons

Baryon is a class of hadron that contains three quarks typified by the proton and neutron, which in my Quark Theory the proton (quarks uud) and neutron (quarks dud) have a structure of quarks inside it where the *down* quark orbits the *up* quark.

So, if nature is built practically on the proton, neutron, and electron, and that inside the proton and neutron the quarks have a structure, then it comes into question if the rest of the baryons shown in Table 2.1 have also a structure? Also, is it possible for a baryon to have a structure if it is made entirely of three *up* quarks or three *down* quarks? That is, there is no *down* quark that orbits a three *up* quarks baryon or there is no up quark where the three *down* quarks can orbit.

Note that the other baryons were the products of high-energy collisions. Accepting the fact that these baryons indeed have the right content of quarks then the following understanding could be entertained: these baryons should be accepted as just what they are and that they could just be a plain product of the collision.

Summary

The current understanding of the quarks inside the proton and neutron is that they have no structure. In this chapter, I had theorized the structure of the quarks inside the proton and neutron with the application of the Charge Theory that explains why a neutron doesn't spin (neutral charge), which in the process supports the existence of quarks inside the proton and neutron.

I had also theorized that there is only the *up* and *down* quark and the rest of their so-called families or generations are actually their *orders* or the *up* and *down* quarks with higher energies.

Chapter 6
Physics

There is nothing more exciting in the forwarding of a new theory than in the application or testing of the new theory to an established knowledge and to use it to make further discoveries. This chapter is the extension of Chapter 4 *New Model of an Atom* to the theory on the structure of the atom of the elements.

The Periodic Table of the Elements

As early as the first physics book I wrote (my theory on light), I had wondered about the arrangement of the atoms in the Periodic Table of the Elements. Even before I had learned of other works on the rearranging of the form of the Periodic Table, I had thought of having a cut-out of the Periodic Table and then connecting the two sides of it to form a cylinder to connect the series of atomic number of the elements with the purpose of somehow making some sense of it. I imagined the Lanthanide Series and Actinide Series forming a loop inside the cylinder. For quite some time I had struggled to understand how the elements were created even before I reached an understanding of the structure of the atom of the elements.

My journey to the understanding of the Periodic Table of the Elements had gone through the evolution of my understanding of the underlying structure of the atom and the existing knowledge in physics. With my discoveries on the structure of the atom of the elements, how far we can master the elements opens up a new world in physics.

Atomic Number Z: Number of Protons and Electrons

The atom of the elements is practically defined by the number of protons (which is equal to the number of electrons) referred to as its atomic number represented by Z.

Z = Number of protons = Number of electrons

My New Model of an Atom supports the Periodic Table of the Elements in that the structure of an atom of the element always *produces* an equal number of protons to electrons as only one electron orbits a proton.

Determining the Number of Neutrons in an Atom

The Periodic Table of the Elements is sequentially arranged according to the number of the protons in an atom. While the Periodic Table of the Elements is sequentially arranged and the placement of each atom of the elements shows the periodic properties of the elements, it should be observed later that the amount of the neutron in each elements also plays a very important factor in the further research to the structure of the atom as the number of neutrons in an atom very much determines the properties and stability (radioactivity) of an atom of an element.

As the Periodic Table only shows the atomic number, atomic mass, and number of proton and electron in an element, the number of neutron can be determined by subtracting the number of protons (denoted by the atomic number, Z) into the atomic mass (denoted by A, which is the total mass of an atom comprising the total number of protons, neutrons, and electrons). Thus,

A = Atomic mass

Z = Number of protons in an atom

Number of neutrons in an atom = A − Z

Note that the mass of an atom is very much contained in the nucleus, which is composed of the protons and neutrons.

(Refer to *Appendix B*, for the atomic number, atomic mass, and the number of protons, and neutrons in an atom of the Periodic Table of the Elements.)

Electron Configuration Pattern

The electron configuration pattern for an atom is given by:

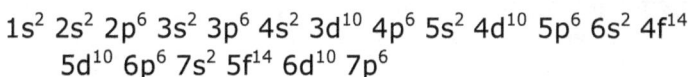

$$1s^2 \; 2s^2 \; 2p^6 \; 3s^2 \; 3p^6 \; 4s^2 \; 3d^{10} \; 4p^6 \; 5s^2 \; 4d^{10} \; 5p^6 \; 6s^2 \; 4f^{14}$$
$$5d^{10} \; 6p^6 \; 7s^2 \; 5f^{14} \; 6d^{10} \; 7p^6$$

(Refer to *Appendix C* for the complete electron configuration of the atoms of the elements in the Periodic Table of the Elements.)

The Periodic Table of the Elements

Figure 6.1 shows the Periodic Table of the Elements. Some designs of the form of the Periodic Table prefer to put the helium atom near the hydrogen atom, while most Periodic Tables put the helium atom in the last column above neon comprising Group VIII of the so-called noble gases. Hydrogen and helium atom has an electron configuration of $1s^1$ and $1s^2$, respectively, while the noble gases have the last term of p^6.

Periodic Table of the Elements

Group																	Noble Gases
1																	18
1A																	8A
1 H	**2** 2A				Transition Elements								**13** 3A	**14** 4A	**15** 5A	**16** 6A	**17** 7A	**2** He
3 Li	**4** Be												**5** B	**6** C	**7** N	**8** O	**9** F	**10** Ne
11 Na	**12** Mg	**3** 3B	**4** 4B	**5** 5B	**6** 6B	**7** 7B	**8**	**9** 8B	**10**	**11** 1B	**12** 2B		**13** Al	**14** Si	**15** P	**16** S	**17** Cl	**18** Ar
19 K	**20** Ca	**21** Sc	**22** Ti	**23** V	**24** Cr	**25** Mn	**26** Fe	**27** Co	**28** Ni	**29** Cu	**30** Zn		**31** Ga	**32** Ge	**33** As	**34** Se	**35** Br	**36** Kr
37 Rb	**38** Sr	**39** Y	**40** Zr	**41** Nb	**42** Mo	**43** Tc	**44** Ru	**45** Rh	**46** Pd	**47** Ag	**48** Cd		**49** In	**50** Sn	**51** Sb	**52** Te	**53** I	**54** Xe
55 Cs	**56** Ba		**72** Hf	**73** Ta	**74** W	**75** Re	**76** Os	**77** Ir	**78** Pt	**79** Au	**80** Hg		**81** Tl	**82** Pb	**83** Bi	**84** Po	**85** At	**86** Rn
87 Fr	**88** Ra		**104** Rf	**105** Db	**106** Sg	**107** Bh	**108** Hs	**109** Mt	**110** Ds	**111** Rg	**112** Cp		**113** Uut	**114** Uuq	**115** Uup	**116** Uuh	**117** Uus	**118** Uuo

Inner Transition Elements

Lanthanide Series

57 La	**58** Ce	**59** Pr	**60** Nd	**61** Pm	**62** Sm	**63** Eu	**64** Gd	**65** Tb	**66** Dy	**67** Ho	**68** Er	**69** Tm	**70** Yb	**71** Lu
89 Ac	**90** Th	**91** Pa	**92** U	**93** Np	**94** Pu	**95** Am	**96** Cm	**97** Bk	**98** Cf	**99** Es	**100** Fm	**101** Md	**102** No	**103** Lr

Actinide Series

Figure 6.1. Periodic Table of the Elements.

Electron Configuration Pattern on the Periodic Table

Looking at the Periodic Table, the use of the atomic number shows the groupings of the properties of the elements. What is not outwardly shown by the Periodic Table is that the last term of the electron configuration shows a pattern that reflects the grouping of the property of the elements (Figure 6.2).

$1s^1$																	$1s^1$
$2s^1$	$2s^2$											$2p^1$	$2p^1$	$2p^1$	$2p^1$	$2p^1$	$2p^1$
$3s^1$	$3s^2$											$3p^1$	$3p^2$	$3p^3$	$3p^4$	$3p^5$	$3p^6$
$4s^1$	$4s^2$	$3d^1$	$3d^2$	$3d^3$	$3d^5$	$3d^5$	$3d^6$	$3d^7$	$3d^8$	$3d^{10}$	$3d^{10}$	$4p^1$	$4p^2$	$4p^3$	$4p^4$	$4p^5$	$4p^6$
$5s^1$	$5s^2$	$4d^1$	$4d^2$	$4d^4$	$4d^5$	$4d^5$	$4d^7$	$4d^8$	$4d^{10}$	$4d^{10}$	$4d^{10}$	$5p^1$	$5p^2$	$5p^3$	$5p^4$	$5p^5$	$5p^6$
$6s^1$	$6s^2$	$5d^1$	$5d^2$	$5d^3$	$5d^4$	$5d^5$	$5d^6$	$5d^7$	$5d^9$	$5d^{10}$	$5d^{10}$	$6p^1$	$6p^2$	$6p^3$	$6p^4$	$6p^5$	$6p^6$
$7s^1$	$7s^2$	$6d^1$	$6d^2$	$6d^3$	$6d^4$	$6d^5$	$6d^6$	$6d^7$	$6d^8$	$6d^{10}$	$6d^{10}$	$7p^1$	$7p^2$	$7p^3$	$7p^4$	$7p^5$	$7p^6$

$4f^1$	$4f^3$	$4f^4$	$4f^5$	$4f^6$	$4f^7$	$1f^7$	$4f^9$	$4f^{10}$	$4f^{11}$	$4f^{12}$	$4f^{13}$	$4f^{14}$	$4f^{14}$
$6d^2$	$5f^2$	$5f^3$	$5f^4$	$5f^6$	$5f^7$	$5f^7$	$5f^9$	$5f^{10}$	$5f^{11}$	$5f^{12}$	$5f^{13}$	$5f^{14}$	$5f^{14}$

Figure 6.2. Electron configuration pattern on the Periodic Table.

Spiral-Circular Periodic Table of the Elements

My early works in the rearrangement of the Periodic Table had formed a cylinder with the Lanthanide and the Actinide Series forming a loop inside the cylinder. As I learned later of some other designs of the Periodic Table, I tried to understand what their arrangement had tried to achieved. It took me awhile of thinking how to take account for the protrusion of the lanthanide and the actinide series before I could come up with my own version of the Periodic Table (Figure 6.3).

I believe that the Spiral-Circular Periodic Table of the Elements is a better representation of the Periodic Table of the Elements since it shows the jumping of the series of elements contained in an energy level to another energy level. The best idea of the design of the Periodic Table is that it suggests that the elements were created in succession. Unless otherwise it will give us a new knowledge of how an atom or the atoms are made, or the structure of the atom of the elements, the original lay-out of the Periodic Table is still a much better arrangement in the observation of the properties of the elements.

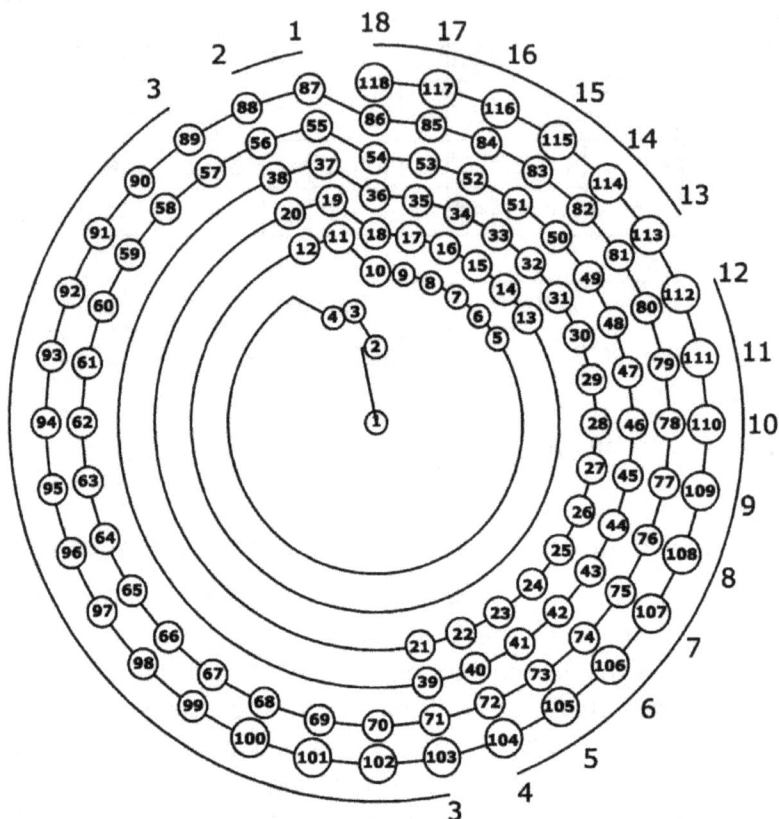

Figure 6.3. Spiral-circular Periodic Table of the Elements.

Theory on the Structure of the Atom of the Elements

In the early times, the alchemists tried to turn lead into gold—literally. Naturally, given their rudimentary understanding of the elements, they failed. In the process, some alchemists in turn turned to the materials that look like gold, such as pyrite, hence the term "false gold." In time, the name alchemy had fallen into disrepute such that its practice became so secretive. One of the most prominent secret alchemists is of course Isaac Newton, who possessed a copy of the Emerald Tablet, which purportedly contained the means to change lead into gold.

Though the search by the alchemists for the means of changing of an element into another was scoffed at, in 1932, John Cockcroft and his collaborator Ernest Walton achieved the first complete artificial nuclear transformation of an element into another. Cockcroft and Walton did it by ripping the electron from the hydrogen atom and accelerating the proton to target the lithium atom. Walton watched in the microscope as the lithium atom changed into helium.

In the evolution of my knowledge, I came into the idea that the elements are *naturally* created in successive fusion of the elements—an observation that can be culled from the Periodic Table of Elements. It could be the successive fusion of the hydrogen atom or of two different elements.

Upon learning of the electron configuration of an atom and my discovery of the New Model of an Atom, I applied this new knowledge to see the structure of an atom. I had chosen the lower atomic number of the oxygen atom for practical purpose of showing the atom. The following are the data for the oxygen:

Oxygen, O
Atomic Number, Z = 8
Proton = 8, Electron = 8, Neutron = 8
Electron configuration: $1s^2\ 2s^2\ 2p^4$

The preliminary plot of the structure of oxygen atom is shown in Figure 6.4.

(I thought earlier that the Pauli Exclusion Principle supports my New Model of an Atom, and vice versa. As I question the possibility of the other quantum numbers, m_l and m_s (which stands for the orbital shape and the spin, respectively) and I am overthrowing the Schrödinger's Model of an Atom with its electron clouds and orbitals then it follows also that I am questioning the validity of the Pauli Exclusion Principle.)

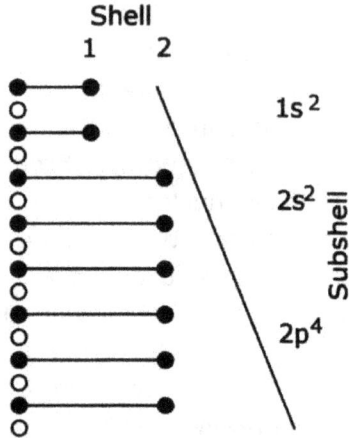

Figure 6.4. The preliminary plot of the structure of the oxygen atom.

Another thing about the plot of the structure of an atom is that in reality, the proton and neutron could be practically very close to each other, possibly in contact with each other, thereby putting all the electrons practically in a flat plane. In this sense, the electrons will repeal each other, arranging them evenly spaced around the nucleus of proton-neutron pole. For example, for a two-electron atom, they will be at $180°$ apart, while that of three-electron atom would be $120°$ (which is derived by dividing $360°$ with the number of electrons in a subshell) bearing in mind the maximum limit of electron for each subshell. Thus the angle of separation angle of the electron for subshell p with the maximum of electrons is $60°$, d with 10 electrons is $36°$, and f with 14 electrons is $25.71°$. To add complexity to the whole arrangement of the electrons in an atom, the electrons from the same shell and other subshells could interact by repulsion with each other to complicate their arrangement.

So, still maintaining the right-angle triangular structure of the atom in Figure 6.4, the atom with the electrons spaced around the nucleus forms into a Christmas tree (Figure 6.5).

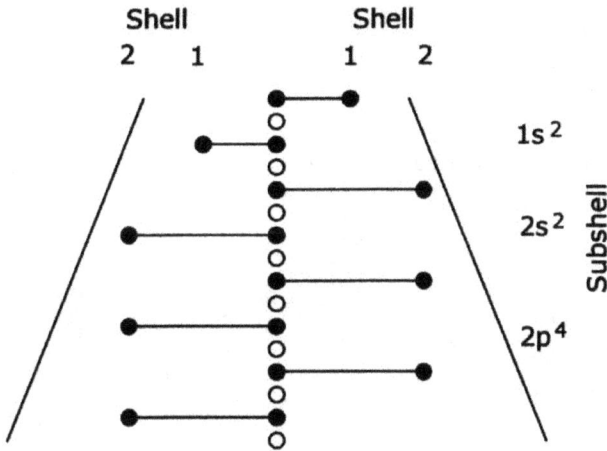

Figure 6.5. The Christmas tree shape of the atom (oxygen) cause by the electrons arranged evenly around the nucleus.

Based from Figure 6.6, the typical depiction of the oxygen atom with two electrons on the first subshell and six electrons on the second subshell could now be understood clearly. What we see as six electrons on the second energy level are actually electrons of the *s* and *p* subshells, and that the electrons are not in the same orbit (see Figure 6.5).

Early in my musings, I had wondered of the result of the nuclear fission of an atom. I had somehow imagined that the element being hit by a "projectile" such as a proton could divide the atom into two and then reforming again into lesser mass atoms. I imagine the structure of the atom to be in the shape of two triangles (Figure 6.7).

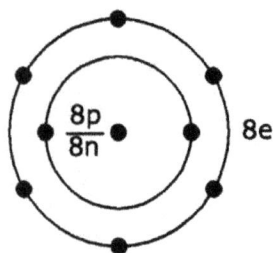

Figure 6.6. My proposed structure of the oxygen atom based on my theories.

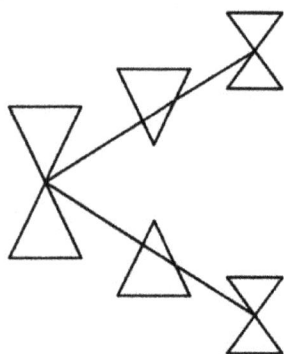

Figure 6.7. My early thoughts on the structure of an atom that splits into two and then reconstituted into another atom of a lighter element.

In trying to observe the arrangement of the atom of the elements in the Periodic Table, I thought of atomic number of the elements following a pattern of going back and forth from side to side (Figure 6.8).

Thinking about this pattern of the going back and forth of the element in Figure 6.7, I applied it on Figure 6.7 (see Figure 6.9).

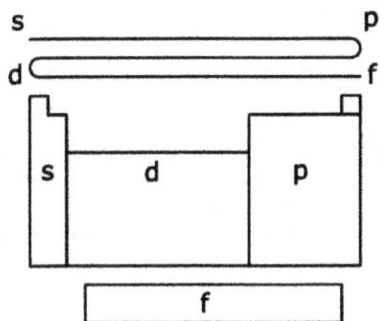

Figure 6.8. My thoughts of the pattern of going back and forth of the element in the Periodic Table.

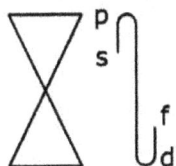

Figure 6.9. The going back and forth pattern in Figure 6.7 applied to the rudimentary structure of an atom.

Remembering the arrangement of the subshell (*s*, *p*, *d*, and *f*) and their position in the Periodic Table (in Figure 6.2), somehow I know I could be on the right track. I returned to the "two triangle" structure of an atom and plot the subshell (Figure 6.10).

The atom of the oxygen cannot show how Figure 6.10 looks like so I used the atom of cerium as it shows electrons on the *d* and *f* subshell but using a shorthand version (Figure 6.11).

Figure 6.10. The subshell plotted on the rudimentary structure of an atom.

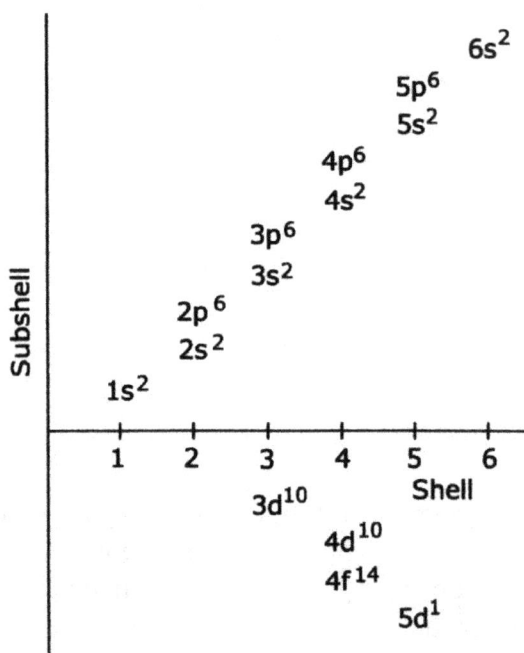

Cerium (Ce), Z=58
Electron configuration pattern:
$1s^2\ 2s^2\ 2p^6\ 3s^2\ 3p^6\ 4s^2\ 3d^{10}\ 4p^6\ 5s^2\ 4d^{10}\ 5p^6\ 6s^2\ 4f^{14}\ 5d^1$

Figure 6.11. The shorthand plot of the atom of cerium showing the separation of the *s* and *p*, and *d* and *f* subshells.

The Extra Neutrons in an Atom

With the proton-electron configuration pattern of an atom, looking at *Appendix B* it will be observed that except for the hydrogen atom, either the element has an equal number of neutron with the proton and electron or that there are extra neutrons in an atom. The question that follows is, "Where do the extra neutrons figure in the structure of an atom?" I am forwarding the following ideas:

Idea #1: Neutrons in excess of the proton-neutron-electron grouping are added at the *end* of the proton-neutron pole of the atom (Figure 6.12).

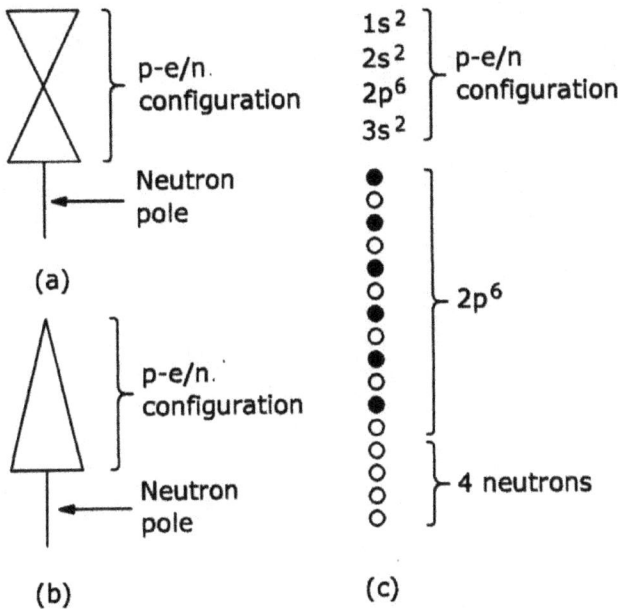

Figure 6.12. Location of the extra neutrons in the proton-neutron pole of an atom. a) On the s/p and d/f shell and subshell configuration. b) On the Christmas tree configuration. c) On the shorthand depiction of an example argon atom with an extra four neutrons shown.

Idea #2: There is a still to be discovered universal pattern of distribution of the neutrons inside the atom. (Note: Idea #2 could use *Appendix B* and *Appendix C* if there is indeed a pattern of arrangement of the neutrons within an atom. Even the isotopes of the atom of the elements may give us an answer.)

Idea #3: The arrangement of the neutrons in an atom is on an individual basis.

Idea #4: Neutrons are randomly moving along or around the proton-neutron pole nucleus (Figure 6.13.).

proton-electron/neutron configuration

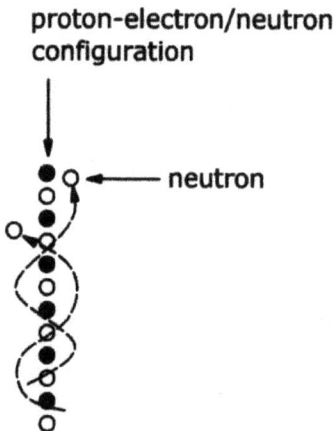

— neutron

Figure 6.13. Neutrons are randomly moving along or around the proton-neutron pole nucleus.

Postulate #5: Neutrons are fixed in place along and around the proton-neutron pole nucleus (Figure 6.14.).

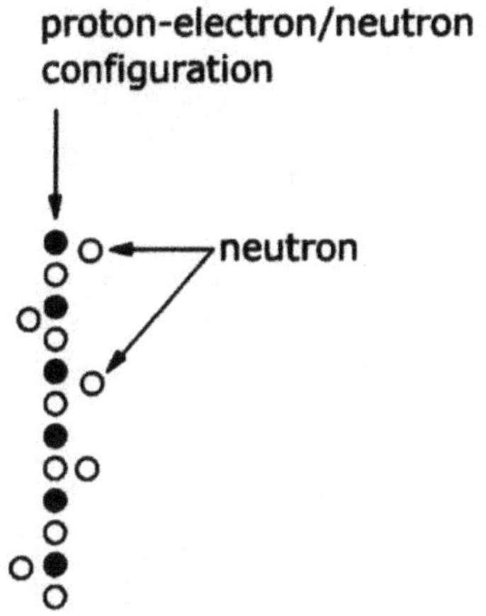

**proton-electron/neutron
configuration**

Figure 6.14. Neutrons are fixed in place along and around the proton-neutron pole nucleus.

In the Idea #1 shown by Figure 6.12, I imagine balancing the "plates" of proton-neutron-electron on a pole of neutron where the extra neutron pole made it unstable or wobbly.

The Size of the Atom of the Elements

Currently, the size of the atom is depicted with the increasing number of electron shells and the nucleus getting larger.

In the New Model of an Atom, the atom of the element is the configuration of the hydrogen configuration (proton and electron) and the neutron stacked on top of each other forming like a dipole magnet. The resultant combination of their electromagnetic fields is the shape and size of the atom (Figure 6.15).

To some extent, the hydrogen and helium atom are in a spherical shape but for the higher elements, they could be in a pear or football shape.

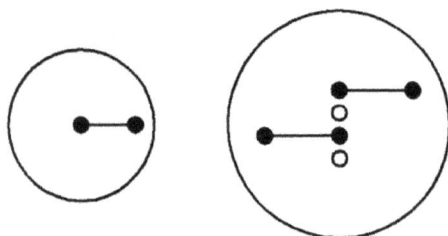

Hydrogen atom   Helium atom

Figure 6.15. Side View: The "size" of the atom is the combined electromagnetic field of the stacked protons and neutrons, and the electrons. Figure viewed on the side view.

Atom Law

The Pauli Exclusion Principle states that no two electrons in an atom can have the same four quantum numbers n, ℓ, m_ℓ and m_s, where these quantum numbers relates to the shell, subshell, the shape of the orbital, and the spin of the electron, respectively. In the New Model of an Atom, the quantum numbers m_ℓ can be eliminated based on the idea that the electron does not travel in the supposed orbital shell of the electron cloud, while that of m_s can be eliminated based on the idea that the electron only physically spin in one direction (clockwise) as its charge.

Based from the New Model of an Atom, a law can be forwarded called Atom Law that governs all the atoms:

Only one electron can orbit a proton.

This law is very much clear, although unrecognized, from the Periodic Table of the Elements, wherein the atomic number, Z, the number of proton is equal to the number of electrons in an atom of the element.

The second law, which I am not totally inclined to push, would be the grouping of the proton, electron, and neutron.

If the Pauli Exclusion Principle pushes for <u>all</u> the quantum number, then it is questionable. If it pushes only for a <u>new</u> definition that no two electrons can be in the same shell (n) and subshell (ℓ), then the Atom Law could practically replace it as the Atom Law is much more clearer in that no two electrons can be in both the same shell and subshell as only one electron orbits a proton.

Summary

Bohr's Model of an Atom, Schrödinger's Model of an Atom, the shell and subshell, Pauli Exclusion Principle, and orbital diagram tried to deal with the structure of the electron of an atom, while the Nuclear Shell Model of an Atom tried to deal with the lump structure of the nucleus of an atom (Figure 6.16). This vagueness in our knowledge of the atom could have warned us that there is something wrong with our model of an atom. As I had observed, even if a theory is questionable and even if somebody recognized that it is wrong but that there is no other idea to replace it, then oftentimes the status quo is maintained. The answer to this confusion in physics is the New Model of an Atom.

The New Model of an Atom is the basic understanding of the atom which shows the charges of the subatomic particles (Charge Theory), how they got their masses (Mass Theory), and the rudimentary structure of the atom with the location of the neutron.

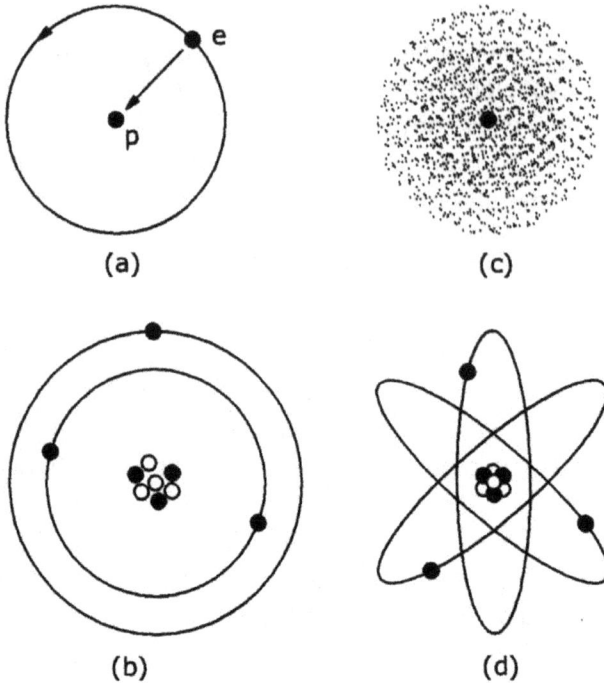

Figure 6.16. From Bohr's Model of an Atom to Schrödinger's Model of an Atom. a) Bohr's Model of an Atom b) Shell and subshell, Pauli Exclusion Principle, and the lump of protons and neutrons in the nucleus c) The electron cloud d) The "simplified" depiction of an atom with the probabilistic motion of the electrons in the orbitals.

Figure 6.17 shows the applications of my theories on the atom of the element that incorporates the Charge Theory, the New Model of an Atom, and the Theory on the Structure of the Atom of the Elements (Figure 6.17).

The New Model of an Atom is the basic description of an atom, once there is an extra neutron involved, this would go to the Theory on the Structure of the Atom of an Element.

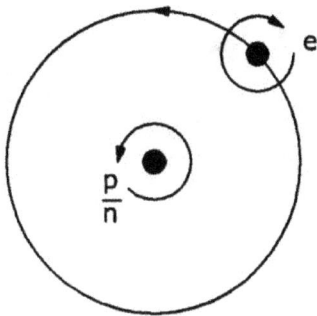

Top view
(a)

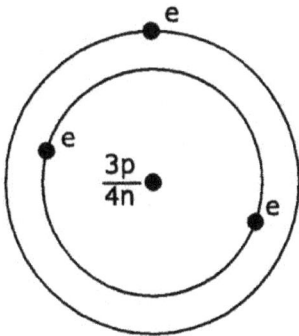

Top view
(b)

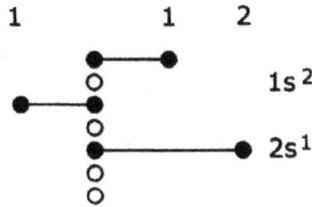

Side view
(c)

Figure 6.17. The proposed model of an atom with its application to the structure of the atom of the elements. a) The New Model of an Atom with the explanation of the charge and the proton as the one that makes the electron orbit around it. b) Top view of the lithium atom with its shell and the number of electrons and the nucleus. c) Side view of the lithium atom showing the shell and subshell and the nucleus.

Chapter 7
Chemistry

As a natural course of this book, chemistry is the next field of application of my theories. Discussions below will reinforce the theory on the model and the structure of the atom.

Dipole Particle

Elementary particles such as electron and quarks inside proton and neutron are like a dipole magnet, spinning on their axis and inducing their energy field, which is the source of their magnetic property (Figure 7.1a). The protons as a nucleus of an atom forms into a configuration of a dipole magnet (Figure 7.1b). The dipole atoms of a bar magnet are aligned to produce the resultant magnetic field (Figure 7.1c).

From the quarks and electron, to the proton and neutron subatomic particles, to the hydrogen atom, to the atom of higher elements illustrates the scaling property of the dipole particle.

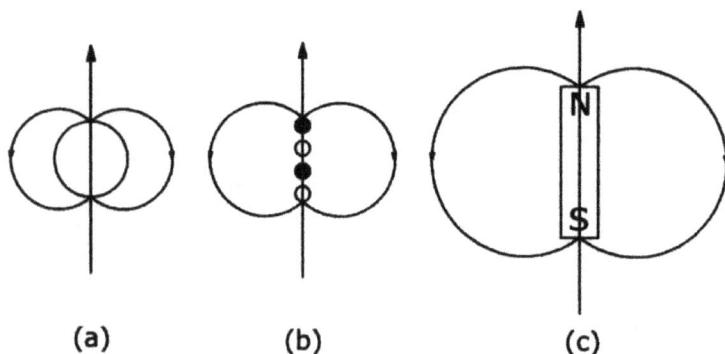

Figure 7.1. Dipole magnet configuration: a) Dipole particle b) Atom structure showing only the nucleus c) The bar magnet and its magnetic field.

The Structure of the Water Molecule

The idea that gave me an insight into the world of molecules was when I had read that in a water molecule there is an angle between the two hydrogen atoms that are attached to the oxygen atom. Based from experimental evidence, the angle between the two hydrogen, referred to as the *bond angle*, was found to be 104.5°.

In what is called the Space-Filling Model of a molecule, the water molecule is typically depicted with no scale to the sizes of its atoms (Figure 7.2).

Taking into account the practical size of the atoms of the hydrogen and oxygen, the oxygen atom would be about eight times the size of the hydrogen atom.

What the Space-Filling Model shows is that the atoms of the molecules are imbedded within each other into much deeper region of their energy fields. Bearing this in mind, in the case of the water molecule, we can say that the only way the hydrogen atoms and the oxygen atom of the water molecule are attached to each other could be by their poles (Figure 7.3).

We can envision the molecule of water by substituting the oxygen atom with a bigger magnet, while that of the two hydrogen atoms with a smaller magnet (Figure 7.4).

In this way, we can actually see based from the water molecule that the bond angle is actually formed from the opposition of the same top poles of the hydrogen atoms.

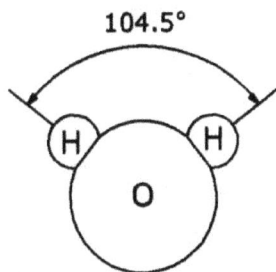

Figure 7.2. The water molecule. Space-Filling Model of a molecule that is not drawn to scale.

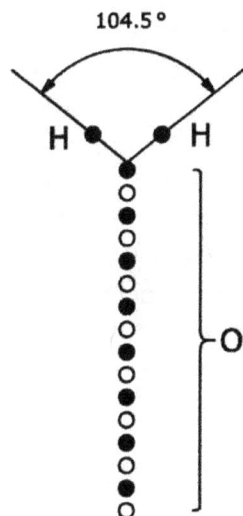

Figure 7.3. The atoms of the water molecule are attached in their poles. (Electrons are not shown.)

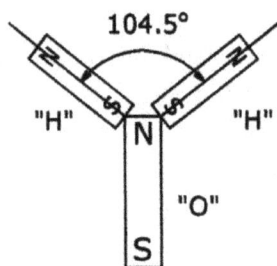

Figure 7.4. The water molecule in terms of bar magnets.

Chemical Bond

What I had discussed above should be reconciled with the concept of chemical bond. (Chemical bond refers to the bonding of the atom of the elements.) According to chemistry books, an atom has core electrons and valence electrons. Core electrons are found near the nucleus of the atom, while the valence electrons are found in the *s* and *p* subshell of the highest energy level. Valence electrons are said to be responsible for holding two or more of atoms together in a chemical bond.

We can say from the above discussions that whereas the formation of the structure of the molecule is based on the dipole structure of the atom, in a chemical bond, the bonding of the atoms is practically governed by the electrons or specifically the valence electrons. That is, it seems to suggest that before two atoms are attached to each other, it has to be permitted or the outcome is governed by the (valence) electrons of the atoms.

Atoms of the Molecule and the Bond Angle

There are two ways of depicting the molecule: one is through the structure formula, and the other is by using the "Ball and Stick" Model. The "Ball and Stick" Model, usually made of plastic material, is consists of a ball representing the atom with pattern of holes around it for attaching the stick. Both the structure formula and the "Ball and Stick" Model (drawn) can best be shown by the molecules of ammonia (NH_3) and phosphine (PH_3), with their bond angle of 107° and 94°, respectively (Figure 7.5).

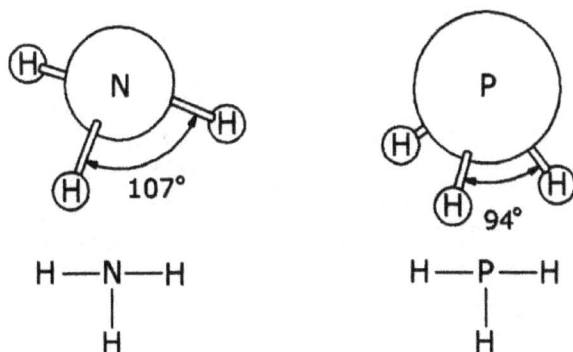

Figure 7.5. The "Ball and Stick" Model of the NH_3 and PH_3 molecules and their structure formula.

Notice that in the "Ball and Stick" Model, the attachment of the hydrogen atoms to the nitrogen and phosphorus are practically anywhere around it. Based from the Space-Filling Model and the atom as a dipole magnet, we can already see that although the "Ball and Stick" Model is an actual practical representation of a molecule, it actually abstract the true understanding of the attachment and formation of the atoms.

Based from the structure of the molecule in Figure 7.5, in applying the dipole magnet model, the hydrogen atoms are practically arranged evenly on its attachment on the nitrogen or phosphorus atom as the hydrogen atoms repel each other evenly (Figure 7.6).

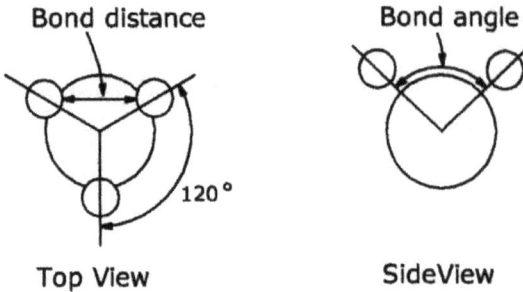

Figure 7.6. Looking from the top of the molecule, the hydrogen atoms are evenly placed at 120° for the three hydrogen atoms (360°/3). The bond angle is the angle that is viewed from the side of the molecule (the figure shows only the angular relation of the two hydrogen atom). We can add another term called "bond distance" to denote the distance of the hydrogen atom to another hydrogen atom.

Based from Figure 7.6, we can derive a conclusion that as the nucleus of the atom increases in its number of protons (and neutrons), the stronger is its magnetic strength. We can see this from the 107° bond angle of ammonia (NH_3) to the 94° bond angle of phosphine (PH_3), where the phosphorous atom has more protons (and neutrons and electrons) than the nitrogen atom. Thus, we can deduce that in the example of the phosphorous atom, the nucleus pulls the hydrogen atoms closer to each other, even as there is repulsion between the hydrogen atoms.

Summary

Any developments in physics will naturally affect the field of chemistry. Even if I had not delved deeper into the field of chemistry, I believed that chemistry will also undergo some changes.

While my discussions here is in support of the structure of an atom of the element having a proton-neutron pole as a nucleus, any observant person will notice that it is more than that—it is actually showing my understanding of gravity (read the last paragraph of section *Molecule of the Atoms and the Bond Angle*). What Figure 7.5 and 7.6 shows is why different areas on Earth have different gravity. Potassium is heavier than nitrogen atom. Water has lesser gravity than land and that land have different gravity due to the different mineral concentrations underground.

Part 3

Real New Physics

Chapter 8
Theory of Everything

Unification of the Fundamental Forces

In physics, the idea of attaining the Theory of Everything is through the unification of all the fundamental forces. The typical approach is through the use of mathematics and through partial unification of two or more fundamental forces with the final objective of the unification of the whole. For example, the Electromagnetic Theory of James Clerk Maxwell was said to be the unification of electricity and magnetism that shows that both are fundamentally the same. Actually, it was not a unification at all of the electricity and magnetism. Rather, the Electromagnetic Theory just explains the relationship of electricity and magnetism. The true understanding of electricity and magnetism is that electricity is the movement of electron, while magnetism is the radiation of electron (and even a free moving particle). Magnetism as a force naturally can influence the electron in as much as the moving electron with its radiation (magnetism) can influence a magnetic force and its source.

Suffice to say, there is really no unification at all of all the fundamental forces. Rather, it is the understanding of what and where those forces are and how they operate. (And for this, physicists had to really think hard if Electroweak Theory is really the unification of the weak force and electromagnetism. As will be shown later down below, they are not even acting on the same area.)

The *New* Fundamental Forces

The current fundamental forces are: electromagnetism, gravity, strong force, and weak force. (The strong force, also called strong interaction and strong nuclear force, is referred to as the one holding the nucleus together in an atom but this is also referred to as the color force when referring to quarks.) From what I had learned, I am proposing for a new classification of the fundamental forces. In the new classification of the fundamental forces, I divided the fundamental forces into two categories: one is that which we can observe in nature and the other is that which pertains to the atom.

The fundamental forces in nature are:

- Electricity or electromagnetism (electron)
- Light (photon)
- Gravity
- Radioactivity

Both electricity and light were before under electromagnetism. Electron and photon are both energy particles and in their state of being free moving particles they emit radiation—of which in the case of electron is called as "electro" magnetism. The weak force (also called weak nuclear force or weak interaction), which is responsible for radioactivity (radioactive decay) of an atom through its neutrons belongs to this category since its effect can be readily seen or felt just like gravity.

The fundamental forces in atom are:

- Atomic weak force
- Atomic strong force
- Quark weak force

- Quark strong force

The atomic weak force in this new fundamental force pertains to the force that binds electron to the proton or the nucleus. It is weak in the sense that the proton forces the electron to orbit around it, yet the electron could be dislodge in its place by photon (light) or another electron (and its electromagnetism). The atomic strong force pertains to the force that holds the nuclei together. In this case, the pion (classified as a meson) is the gauge boson or the particle that transmits this force (Figure 8.1).

Atomic Weak Force: $W^{\pm}$, Z^0

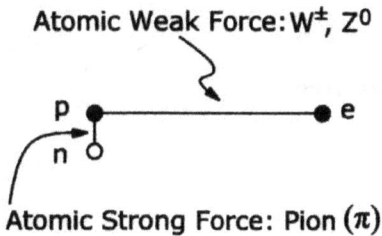

p ●————————● e

n

Atomic Strong Force: Pion (π)

Figure 8.1. The atomic weak force and strong force in an atom. (The atom is deuterium, which an isotope of hydrogen atom.)

What is called previously as the strong force that pertains to the quarks is also called as the color force. (Quarks are not observed outside of the protons and neutrons.) That is, since I classified a quark strong force and quark weak force, compared to the atomic strong force, the quark weak force is obviously much stronger. Based from my Quark Theory, which forwards the structure of the quarks inside the proton and neutron (refer to Figure 5.1), Figure 8.2 shows where the quark strong force and quark weak force are within the proton and neutron.

The dominant theory on quarks right now is the Quantum Chromodynamics (QCD), which describes the interactions of the quarks and gluons inside the hadrons such as proton and neutron. I had always felt an unsettling feeling on the use of the term "color," as I felt that "color" is like as unreal as the quantum number pertaining to the spin of a particle. Surely, physicists will have to deal with these questions as it is beyond my knowledge, work, and attention to pursue these questions. If nothing else, the abstract use of "color" of QCD only signals an underlying need of more plausible explanation of the quarks.

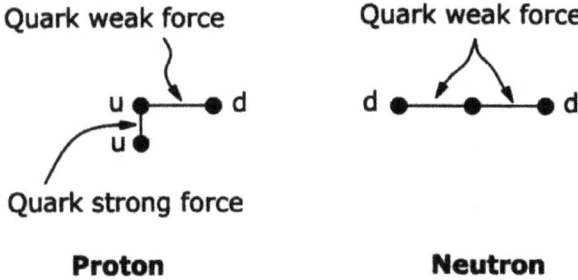

Proton   **Neutron**

Figure 8.2. The quark weak force and quark strong force within the proton and neutron.

The *New* Fundamental Particles

The first time I learned of the Standard Model, I just accepted it as it is. I thought that it is beyond my ability to understand the Standard Model as it is the product of experiments and discoveries requiring specialized scientific knowledge and machines. As luck would have it, I am rather working on the works of the physicists that requires nothing but seeing what they had accomplished on a different point of view.

My work on the Standard Model could now be partially described to show that the *up* and *down* quark as the fundamental fermion particles, while that of the lepton is the electron. The other quarks under the *up* and *down* quarks are their orders or their higher-energy self, just as the other leptons under the electron are its orders (Figure 8.3).

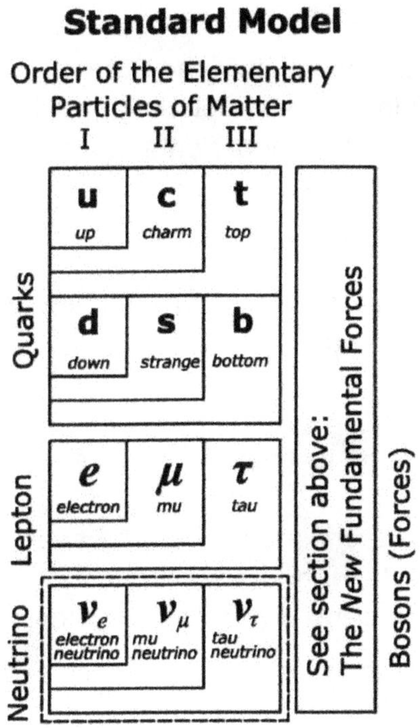

Standard Model

Order of the Elementary
Particles of Matter

Figure 8.3. Table of the orders of quarks and leptons.

What Figure 8.3 shows is that there could still be more orders of quarks with higher masses that could be discovered using very high particle accelerator. (The question is—at what price we want to pay for these discoveries just to prove that they exist? Though it cannot be denied, the triumph of the Standard Model was done by the hard work of the physicists who had discovered these particles for which I just simply and significantly understood them to be what they are. Without their works, I could not have accomplished my work.)

What Figure 8.3 with a dash lines on the neutrinos shows is that matter is only made up of the *up* and *down* quarks, and electron lepton; the neutrinos should be included into the fundamental forces related to radioactivity.

THEORY OF EVERYTHING

$E=mc^2$ and the Standard Model

The Standard Model being a candidate for the Theory of Everything needs $E=mc^2$ to explain the mass of the particles. By themselves alone, they both lack the "ingredients" to be the Theory of Everything. (Charge Theory should be included to the Standard Model and $E=mc^2$ to attain the basic ingredients of the Theory of Everything.)

For $E=mc^2$, it is an incomplete general description of everything in the universe. The equation is broken down as follows:

- E. Everything in the universe is energy, as in energy particle.
- c. Speed or motion is an inherent property of the energy particle. No particle can exists without moving.
- m. Mass is the derived inherent property of a moving energy particle, $m=E/c^2$.

For the Standard Model, it serves as the detailed identification of the energy particles of the universe—its building blocks. The concept of the Standard Model is that the universe is consists of fundamental particles and forces, and that these forces are mediated by particles called bosons. The following are the fundamental particles:

- Quarks: up and down. The energy particle producing the proton and neutron.
- Lepton: electron. The energy particle orbiting the proton. The "workhorse" of matter.

(To really delve deeper into the nature of matter, the quarks and electrons are practically eternal spinning flame—a wonder of wonders if we think of the particles as a fire to eventually die out or even much spin without reason or cause.)

In my view of the strictest understanding of a fundamental particle, the neutrino is not included as a fundamental particle

142

as it is the product of reactions such as the fusion of the atom and the decay (changing back) of the neutron to proton.

Theory of Everything

Can the Theory of Everything be printed in a t-shirt? That is, can it be put into a complete mathematical expression? Well, I'm not going to stick out my head on that; I am not a physicist nor a mathematician to work it all up but I can work it up the best I know—the only thing I know.

As a Theory of Everything, this theory can supposedly explain the universe and everything in it. The constituents of the Theory of Everything are:

Fundamental particles. Everything in the universe is an energy particle ($E=mc^2$). The particle of matter is consists of quarks (up and down) and lepton (electron)—the Standard Model. (Philosophically, it has to be this simple to simplify the creation of the complex universe, much like the 0 and 1 of the computer can create the complexity of the virtual world of the computer or its ability to express practically everything in our physical world.)

Fundamental forces. Fundamental forces answers the following questions:

- What holds the quarks in the proton and neutron?
- What holds the subatomic particles in the atom together?
- What are these forces?
- Where are these forces? How do we experience these forces?
- How are these forces given off?

Fundamental properties. Spin (Charge Theory), mass ($m=E/c2$), speed (travel, orbit, and spin of the particle) and field (dipole particle).

Structure. The structure of the quark inside the proton and neutron (Quark Theory). The structure of the proton, neutron, and electron inside the atom (New Model of an Atom and Theory of the Structure of the Atom of the Elements).

Operation and Laws. Knowing the constituents of the Theory of Everything, the laws and how nature operates or its mechanism can be known or discovered. The following are just the few of the discovered laws of nature:

- $E=mc^2$. Everything in the universe is made of energy particle, which are in motion and by virtue of its energy and motion, the particle attains a mass.
- Atom Law. The law states that only one electron can orbit a proton.
- Charge Theory. The spin of a particle: clockwise spin as positive charge, counterclockwise spin as negative charge, and non-spin of a neutral charge.
- Periodic Table of the Elements. The number of proton, neutron, and electron in an atom; the discovery of the shell and subshell of the electron; the maximum electrons in the subshells; the electron configuration pattern of the atom of the elements; the periodicity of the properties of the elements (Periodic Table of the Elements); etc...

Part of my Theory of Everything that will still be published is my theory on light and another book that is still to be written is my theory on gravity (or rather a problem that I will still solve).

Reorganizing Physics

Physics is currently divided into two fields of study: Classical Physics and Quantum Physics. Classical Physics, which is also called Classical Mechanics or just Mechanics deals with the study of the behavior of physical bodies when subjected to motion or displacement and their subsequent effect on their environment. That is, Classical Physics deals with the everyday world we see or experience. Quantum Physics, which is also called Quantum Mechanics or sometimes Quantum Theory, deals with the physical phenomena at microscopic scales.

The current condition in physics is that Classical Physics and Quantum Mechanics do not agree with each other. For example, Quantum Mechanics cannot explain gravity and the universe, while Classical Physics with Einstein's General Theory of Relativity that explains gravity cannot explain the minute world of the atom. A theory called Quantum Gravity Theory is still being searched by the physicists to finally merge General Relativity Theory and the Quantum Theory.

Time is very near where physics will be a seamless field of the atom and the whole universe. (This is in fact the subject of books and discussions of the particle physicists and astrophysicists about the explanation of the universe from quarks to cosmos.) That is, while Classical Physics and Quantum Mechanics are still two fields of study, the idea of the barrier of the small world of the atom and larger world of the universe will disappear and they will become one continuous subject.

Epilogue

Some science writers and physicists think that great discoveries in physics, particularly in the field of particle physics, cannot be done anymore by a single person but by groups of physicists connected with an academe or in scientific establishments using big machines. So here I am—the proverbial "lone wolf." I will have vested many physicists on the many researches related to the subjects of this book not because I am brilliant—rather that I am lucky and that I had used my intuition to guide me along the way. Being an outsider in the field of physics had worked to my advantage as I was free from the pressures of following a well-treaded path such as those imposed by the academe, research facilities, and the practicality of practicing physics as a means of earning a living.

The question on everyone's mind right now that begs to be answered would be: "What's going to happen next?" What I had covered in this book would probably occupy a lot of people into filling up the details. The search for the Higgs boson will be settled out. Physicists will settle the need for the big machines. Charge and spin will be reconciled as one and the same, banishing spin as just "quantum number" and instead recognize it as a real property of the particles. The New Model of an Atom will be tested and its mastery will be pursued to the fullest.

The future of my work, that is, the publication of my other books will largely depend on this book and how it will be received. I guess I am just being realistic—actually pessimistic.

Suffice to say, based from my new understandings, I will still make revisions to my first book, theory of light. I will still have to write (solve) my second book, my theory on gravity, which is the long-sought Quantum Gravity Theory. It will still take me a long time and the usual hard labor to finish what I had set out to do—probably three to four years. I will try my best to publish all my books.

Have I brought down or weaken the support for the String Theory that would have release the time, money, and brain power being spent on it, then my next attention is to bring down the Big Bang Theory. The Big Bang Theory is a very formidable theory to bring down considering the immense works and support for it—and that makes it more challenging and exciting to overthrow. I may not do a direct attack on the Big Bang Theory but rather present my theory or theories so much that all that was attributed to the Big Bang Theory will be explained by my theory or theories. That is, I will drain the Big Bang Theory of its strength and support.

My works will affect both the living and the dead who had received the Nobel Prize. The thing that I had learned in science is that truth is such an unforgiving master and that applies to my work too. I, myself will never hold on to any of my idea that cannot pass reasonable scrutiny.

If my work does what it is supposed to do—usher in the Real New Physics (Renewal Physics), then we are in for a very exciting time. Physics will need the effort of a lot of people to bring it to fruition. So much will be freed from our minds that addressing the more urgent present problems that will create more jobs and open up other fields of research that really and practically deserve the attention they need: alternative fuel and energy source, improve space propulsion energy source, and the rendering to safety of spent nuclear fuel, among others. The age of the Real New Physics is about addressing the practical and necessary needs of humankind and not the flight of fancy and escapism that the scientific world is in right now.

We are at a major crossroad of scientific achievements and we have the decision on which road to take.

(Can we reach the "Golden Age" of science and address our political, economic, environmental, and sociological problems? Is the current scientific generation, who has the opportunity to do beyond what is the current status quo, is being apolitical and does best by denial and evasion of the decisive path physics should undertake?   Will the future generations say that we played the fiddle while the world is burning?)

Appendix A

Alphabetical Listings of the Elements by Chemical Name

Alphabetical List of the Elements

Element	Symbol	Atomic Number
Actinium	Ac	89
Aluminum	Al	13
Americium	Am	95
Antimony	Sb	51
Argon	Ar	18
Arsenic	As	33
Astatine	At	85
Barium	Ba	56
Berkelium	Bk	97
Bismuth	Be	83
Bohrium	Be	4
Boron	B	5
Bromine	Br	35
Cadmium	Cd	48
Calcium	Ca	20
Californium	Cf	98
Carbon	C	6
Cerium	Ce	58
Cesium	Cs	55
Chlorine	Cl	17
Chromium	Cr	24

Cobalt	Co	27
Copernicium	Cn	112
Copper	Cu	29
Curium	Cm	96
Darmstadtium	Ds	110
Dubnium	Db	105
Dysprosium	Dy	66
Einsteinium	Es	99
Erbium	Er	68
Europium	Eu	63
Fermium	Fm	100
Flerovium	Fl	114
Fluorine	F	9
Francium	Fr	87
Gadolinium	Gd	64
Gallium	Ga	31
Germanium	Ge	32
Gold	Au	79
Hafnium	Hf	72
Hassium	Hs	108
Helium	He	2
Holmium	Ho	67
Hydrogen	H	1
Indium	In	49
Iodine	I	53
Iridium	Ir	77
Iron	Fe	26
Krypton	Kr	36
Lanthanium	La	57
Lawrencium	Lr	103
Lead	Pb	82
Lithium	Li	3
Livermorium	Lv	116
Lutetium	Lu	71
Magnesium	Mg	12
Manganese	Mn	25
Meitnerium	Mt	109
Mendelevium	Md	101

Mercury	Hg	80
Molybdenum	Mo	42
Neodymium	Nd	60
Neon	Ne	10
Neptunium	Np	93
Nickel	Ni	28
Niobium	Nb	41
Nitrogen	N	7
Nobelium	No	102
Osmium	Os	76
Oxygen	O	8
Palladium	Pd	46
Phosphorous	P	15
Platinum	Pt	78
Plutonium	Pu	94
Polonium	Po	84
Potassium	K	19
Praseodymium	Pr	59
Promethium	Pm	61
Protactinium	Pa	91
Radium	Ra	88
Radon	Rn	86
Rhenium	Re	75
Rhodium	Rh	45
Roentgenium	Rg	111
Rubidium	Rb	37
Ruthenium	Ru	44
Rutherfordium	Rf	104
Samarium	Sm	62
Scandium	Sc	21
Seaborgium	Sg	106
Selenium	Se	34
Silicon	Si	14
Silver	Ag	47
Sodium	Na	11
Strontium	Sr	38
Sulfur	S	16
Tantalum	Ta	73

Technetium	Tc	43
Tellurium	Te	52
Terbium	Tb	65
Thallium	Tl	81
Thorium	Th	90
Thulium	Tm	69
Tin	Sn	50
Titanium	Ti	22
Tungsten	W	74
Ununoctium	Uuo	118
Ununpentium	Uup	115
Ununseptium	Uus	117
Uranium	U	92
Vanadium	V	23
Xenon	Xe	54
Ytterbium	Yb	70
Yttrium	Y	39
Zinc	Z	30
Zirconium	Zr	40

Appendix B

Periodic Table of the Elements: Atomic Number; Atomic Mass; and Number of Protons, Electrons, and Neutrons of an Atom

As I forward a New Model of an Atom and the Theory on the Structure of the Atom of the Element, this *Appendix B* in conjunction with *Appendix C* (which is the electron configuration pattern for the elements) is a means to be used as a study in the possible discovery of a universal pattern of the structure of the Periodic Table of the Elements.

Symbol:
- Z   Atomic number
- A   Atomic mass
- P   Number of proton in an atom
- E   Number of electron in an atom
- N   Number of neutron in an atom
- Nx Extra neutron

Derivation:
$$Z = P = E$$
$$N = A - Z$$
$$Nx = N - P$$

Atomic Number, Atomic Mass, and the Number of Protons, Electrons, Neutrons, and Extra Neutrons

Symbol Name of Element		Z	A	P/E	N	Nx
H	Hydrogen	1	1	1	0	0
He	Helium	2	4	2	2	0
Li	Lithium	3	7	3	4	1
Be	Beryllium	4	9	4	5	1
B	Boron	5	11	5	6	1
C	Carbon	6	12	6	6	0
N	Nitrogen	7	14	7	7	0
O	Oxygen	8	16	8	8	0
F	Fluorine	9	19	9	10	1
Ne	Neon	10	20	10	10	0
Na	Sodium	11	23	11	12	1
Mg	Magnesium	12	24	12	12	0
Al	Aluminum	13	27	13	14	1
Si	Silicon	14	28	14	14	0
P	Phosphorous	15	31	15	16	1
S	Sulfur	16	32	16	16	0
Cl	Chlorine	17	35	17	18	1
Ar	Argon	18	40	18	22	4
K	Potassium	19	30	19	21	3
Ca	Calcium	20	40	20	20	0
Sc	Scandium	21	45	21	24	3
Ti	Titanium	22	48	22	26	4
V	Vanadium	23	51	23	28	5
Cr	Chromium	24	52	24	28	4
Mn	Manganese	25	55	25	30	5
Fe	Iron	26	56	26	30	4
Co	Cobalt	27	58	27	31	4
Ni	Nickel	28	58	28	30	2
Cu	Copper	29	64	29	35	6

Zn	Zinc	30	65	30	35	5
Ga	Gallium	31	70	31	39	8
Ge	Germanium	32	73	32	41	9
As	Arsenic	33	75	33	42	9
Se	Selenium	34	79	34	50	6
Br	Bromine	35	80	35	45	10
Kr	Krypton	36	84	36	48	12
Rb	Rubidium	37	85	37	48	11
Sr	Strontium	38	88	38	50	12
Y	Yttrium	39	89	39	50	11
Zr	Zirconium	40	91	40	51	11
Nb	Niobium	41	93	41	52	11
Mo	Molybdenum	42	96	42	54	12
Tc	Technetium	43	98	43	55	12
Ru	Ruthenium	44	101	44	57	13
Rh	Rhodium	45	103	45	58	13
Pd	Palladium	46	106	46	60	14
Ag	Silver	47	108	47	61	14
Cd	Cadmium	48	112	48	64	16
In	Indium	49	115	49	66	17
Sn	Tin	50	119	50	69	19
Sb	Antimony	51	122	51	71	20
Te	Tellurium	52	128	52	76	24
I	Iodine	53	127	53	74	21
Xe	Xenon	54	131	54	77	23
Cs	Cesium	55	133	55	78	23
Ba	Barium	56	137	56	81	25
La	Lanthanum	57	139	57	82	25
Ce	Cerium	58	140	58	82	24
Pr	Praseodymium	59	141	59	82	23
Nd	Neodymium	60	144	60	84	24
Pm	Promethium	61	145	61	84	23
Sm	Samarium	62	150	62	88	26
Eu	Europium	63	152	63	89	26
Gd	Gadolinium	64	157	64	93	29
Tb	Terbium	65	159	65	94	29

Dy	Dysprosium	66	163	66	97	31
Ho	Holmium	67	165	67	98	31
Er	Erbium	68	167	68	99	31
Tm	Thulium	69	169	69	100	31
Yb	Ytterbium	70	173	70	103	33
Lu	Lutetium	71	175	71	104	33
Hf	Hafnium	72	178	72	106	34
Ta	Tantalum	73	181	73	108	35
W	Tungsten	74	184	74	110	36
Re	Rhenium	75	186	75	111	36
Os	Osmium	76	190	76	114	38
Ir	Iridium	77	192	77	115	39
Pt	Platinum	78	195	78	117	39
Au	Gold	79	197	79	118	39
Hg	Mercury	80	201	80	121	41
Tl	Thallium	81	204	81	123	42
Pb	Lead	82	207	82	125	43
Bi	Bismuth	83	209	83	126	43
Po	Polonium	84	209	84	125	41
As	Astatine	85	210	85	125	41
Rn	Radon	86	222	86	136	50
Fr	Francium	87	223	87	136	49
Ra	Radium	88	226	88	138	50
Ac	Actinium	89	227	89	138	49
Th	Thorium	90	232	90	142	52
Pa	Protactinium	91	231	91	140	49
U	Uranium	92	238	92	146	54
Np	Neptunium	93	237	93	144	51
Pu	Plutonium	94	244	94	150	56
Am	Americium	95	243	95	148	53
Cm	Curium	96	247	96	151	55
Be	Berkelium	97	247	97	150	53
Cf	Californium	98	251	98	153	55
Es	Einsteinium	99	252	99	153	54
Fm	Fermium	100	257	100	157	57
Md	Mendelevium	101	258	101	157	56

No	Nobelium	102 259	102	157	54
Lr	Lawrencium	103 262	103	159	56
Rf	Rutherfordium	104 261	104	157	53
Db	Dubnium	105 268	105	N/A	N/A
Sg	Seaborgium	106 263	106	157	51
Bh	Bohrium	107 264	107	157	50
Hs	Hassium	108 269	108	161	53
Mt	Meitnerium	109 268	109	159	50
Ds	Darmstadtium	110 272	110	162	52
Rg	Roentgenium	111 273	111	162	51
Uub	Ununbium	112 277	112	165	53
Uut	Ununtrium	113 286	113	173	60
Uuq	Ununquadium	114 289	114	175	61
Uup	Ununpendium	115 288	115	173	58
Uuh	Ununhexium	116 292	116	176	60
Uus	Ununseptium	117 292	117	175	58
Uuo	Unonoctium	118 293	118	175	57

Appendix C

Electron Configuration of the Atoms of the Periodic Table of the Elements

Note: The bold Electron Configuration Pattern in the square bracket is the complete electron configuration pattern of that element of which the rest below it are the short hand Electron Configuration Pattern of the Elements is added to.

There are 18 elements that are exceptions to the electron configuration for the atom: chromium (24), copper (29), niobium (41), molybdenum (42), ruthenium (44), rhodium (45), palladium (46), silver (47), lanthanum (57), cerium (58), gadolinium (64), gold (79), actinium (89), thorium (90), protactinium (91), uranium (92), neptunium (93), and curium (96).

Electron Configuration Pattern: $1s^2\ 2s^2\ 2p^6\ 3s^2\ 3p^6\ 4s^2\ 3d^{10}$ $4p^6\ 5s^2\ 4d^{10}\ 5p^6\ 6s^2\ 4f^{14}\ 5d^{10}\ 6p^6\ 7s^2\ 5f^{14}\ 6d^{10}\ 7p^6$

Ground State Electron Configuration of All Atoms of the Periodic Table of the Elements

Atomic Number & Element Name	Electron Configuration
1 Hydrogen	$1s^1$

2   Helium $1s^2$
3   Lithium $1s^2\ 2s^1$
4   Beryllium $1s^2\ 2s^2$
5   Boron $1s^2\ 2s^2\ 2p^1$
6   Carbon $1s^2\ 2s^2\ 2p^2$
7   Nitrogen $1s^2\ 2s^2\ 2p^3$
8   Oxygen $1s^2\ 2s^2\ 2p^4$
9   Fluorine $1s^2\ 2s^2\ 2p^5$
10 Neon $1s^2\ 2s^2\ 2p^6$
11 Sodium $1s^2\ 2s^2\ 2p^6\ 3s^1$
12 Magnesium $1s^2\ 2s^2\ 2p^6\ 3s^2$
13 Aluminum $1s^2\ 2s^2\ 2p6\ 3s^2\ 3p^1$
14 Silicon $1s^2\ 2s^2\ 2p6\ 3s^2\ 3p^2$
15 Phosphorous $1s^2\ 2s^2\ 2p^6\ 3s^2\ 3p^3$
16 Sulfur $1s^2\ 2s^2\ 2p^6\ 3s^2\ 3p^4$
17 Chlorine $1s^2\ 2s^2\ 2p^6\ 3s^2\ 3p^5$
18 Argon $1s^2\ 2s^2\ 2p^6\ 3s^2\ 3p^6$
19 Potassium $1s^2\ 2s^2\ 2p^6\ 3s^2\ 3p^6\ 4s^1$
20 Calcium $1s^2\ 2s^2\ 2p^6\ 3s^2\ 3p^6\ 4s^2$
21 Scandium $1s^2\ 2s^2\ 2p^6\ 3s^2\ 3p^6\ 4s^2\ 3d^1$
22 Titanium $1s^2\ 2s^2\ 2p^6\ 3s^2\ 3p^6\ 4s^2\ 3d^2$
23 Vanadium $1s^2\ 2s^2\ 2p^6\ 3s^2\ 3p^6\ 4s^2\ 3d^3$
24 Chromium $1s^2\ 2s^2\ 2p^6\ 3s^2\ 3p^6\ 4s^1\ 3d^5$
25 Manganese $1s^2\ 2s^2\ 2p^6\ 3s^2\ 3p^6\ 4s^2\ 3d^5$
26 Iron $1s^2\ 2s^2\ 2p^6\ 3s^2\ 3p^6\ 4s^2\ 3d^6$
27 Cobalt $1s^2\ 2s^2\ 2p^6\ 3s^2\ 3p^6\ 4s^2\ 3d^7$
28 Nickel $1s^2\ 2s^2\ 2p^6\ 3s^2\ 3p^6\ 4s^2\ 3d^8$
29 Copper $1s^2\ 2s^2\ 2p^6\ 3s^2\ 3p^6\ 4s^1\ 3d^{10}$
30 Zinc $1s^2\ 2s^2\ 2p^6\ 3s^2\ 3p^6\ 4s^2\ 3d^{10}$
31 Gallium $1s^2\ 2s^2\ 2p^6\ 3s^2\ 3p^6\ 4s^2\ 3d^{10}\ 4p^1$
32 Germanium $1s^2\ 2s^2\ 2p^6\ 3s^2\ 3p^6\ 4s^2\ 3d^{10}\ 4p^2$
33 Arsenic $1s^2\ 2s^2\ 2p^6\ 3s^2\ 3p^6\ 4s^2\ 3d^{10}\ 4p^3$
34 Selenium $1s^2\ 2s^2\ 2p^6\ 3s^2\ 3p^6\ 4s^2\ 3d^{10}\ 4p^4$
35 Bromine $1s^2\ 2s^2\ 2p^6\ 3s^2\ 3p^6\ 4s^2\ 3d^{10}\ 4p^5$
36 Krypton $1s^2\ 2s^2\ 2p^6\ 3s^2\ 3p^6\ 4s^2\ 3d^{10}\ 4p^6$
[Kr $1s^2\ 2s^2\ 2p^6\ 3s^2\ 3p^6\ 4s^2\ 3d^{10}\ 4p^6$]
37 Rubidium   [Kr] $5s^1$

38	Strontium	[Kr] $5s^2$
39	Yttrium	[Kr] $5s^2$ $4d^1$
40	Zirconium	[Kr] $5s^2$ $4d^2$
41	Niobium	[Kr] $5s^1$ $4d^4$
42	Molybdenum	[Kr] $5s^1$ $4d^5$
43	Technetium	[Kr] $5s^1$ $4d^6$
44	Ruthenium	[Kr] $5s^2$ $4d^7$
45	Rhodium	[Kr] $5s^1$ $4d^8$
46	Palladium	[Kr] $5s^0$ $4d^{10}$
47	Silver	[Kr] $5s^1$ $4d^{10}$
48	Cadmium	[Kr] $5s^2$ $4d^{10}$
49	Indium	[Kr] $5s^2$ $4d^{10}$ $5p^1$
50	Tin	[Kr] $5s^2$ $4d^{10}$ $5p^2$
51	Antimony	[Kr] $5s^2$ $4d^{10}$ $5p^3$
52	Tellurium	[Kr] $5s^2$ $4d^{10}$ $5p^4$
53	Iodine	[Kr] $5s^2$ $4d^{10}$ $5p^5$
54	Xenon	[Kr] $5s^2$ $4d^{10}$ $5p^6$

[Xe $1s^2$ $2s^2$ $2p^6$ $3s^2$ $3p^6$ $4s^2$ $3d^{10}$ $4p^6$ $5s^2$ $4d^{10}$ $5p^6$]

55	Cesium	[Xe] $6s^1$
56	Barium	[Xe] $6s^2$
57	Lanthanum	[Xe] $6s^2$ $5d^1$
58	Cerium	[Xe] $6s^2$ $4f^1$ $5d^1$
59	Praseodymium	[Xe] $6s^2$ $4f^3$ $5d^0$
60	Neodymium	[Xe] $6s^2$ $4f^4$ $5d^0$
61	Promethium	[Xe] $6s^2$ $4f^5$ $5d^0$
62	Samarium	[Xe] $6s^2$ $4f^6$ $5d^0$
63	Europium	[Xe] $6s^2$ $4f^7$ $5d^0$
64	Gadolinium	[Xe] $6s^2$ $4f^7$ $5d^1$
65	Terbium	[Xe] $6s^2$ $4f^9$ $5d^0$
66	Dysprosium	[Xe] $6s^2$ $4f^{10}$ $5d^0$
67	Holmium	[Xe] $6s^2$ $4f^{11}$ $5d^0$
68	Erbium	[Xe] $6s^2$ $4f^{12}$ $5d^0$
69	Thulium	[Xe] $6s^2$ $4f^{13}$ $5d^0$
70	Ytterbium	[Xe] $6s^2$ $4f^{14}$ $5d^0$
71	Lutetium	[Xe] $6s^2$ $4f^{14}$ $5d^1$
72	Hafnium	[Xe] $6s^2$ $4f^{14}$ $5d^2$

73	Tantalum	[Xe] $6s^2$ $4f^{14}$ $5d^3$
74	Tungsten	[Xe] $6s^2$ $4f^{14}$ $5d^4$
75	Rhenium	[Xe] $6s^2$ $4f^{14}$ $5d^5$
76	Osmium	[Xe] $6s^2$ $4f^{14}$ $5d^6$
77	Iridium	[Xe] $6s^2$ $4f^{14}$ $5d^7$
78	Platinum	[Xe] $6s^1$ $4f^{14}$ $5d^9$
79	Gold	[Xe] $6s^1$ $4f^{14}$ $5d^{10}$
80	Mercury	[Xe] $6s^2$ $4f^{14}$ $5d^{10}$
81	Thallium	[Xe] $6s^2$ $4f^{14}$ $5d^{10}$ $6p^1$
82	Lead	[Xe] $6s^2$ $4f^{14}$ $5d^{10}$ $6p^2$
83	Bismuth	[Xe] $6s^2$ $4f^{14}$ $5d^{10}$ $6p^3$
84	Polonium	[Xe] $6s^2$ $4f^{14}$ $5d^{10}$ $6p^4$
85	Astatine	[Xe] $6s^2$ $4f^{14}$ $5d^{10}$ $6p^5$
86	Radon	[Xe] $6s^2$ $4f^{14}$ $5d^{10}$ $6p^6$

[Rn $1s^2$ $2s^2$ $2p^6$ $3s^2$ $3p^6$ $4s^2$ $3d^{10}$ $4p^6$ $5s^2$ $4d^{10}$ $5p^6$ $6s^2$ $4f^{14}$ $5d^{10}$ $6p^6$]

87	Francium	[Rn] $7s^1$
88	Radium	[Rn] $7s^2$
89	Actinium	[Rn] $7s^2$ $6d^1$
90	Thorium	[Rn] $7s^2$ $5f^0$ $6d^2$
91	Protactinium	[Rn] $7s^2$ $5f^2$ $6d^1$
92	Uranium	[Rn] $7s^2$ $5f^3$ $6d^1$
93	Neptunium	[Rn] $7s^2$ $5f^4$ $6d^1$
94	Plutonium	[Rn] $7s^2$ $5f^6$ $6d^0$
95	Americium	[Rn] $7s^2$ $5f^7$ $6d^0$
96	Curium	[Rn] $7s^2$ $5f^7$ $6d^1$
97	Berkelium	[Rn] $7s^2$ $5f^9$ $6d^0$
98	Californium	[Rn] $7s^2$ $5f^{10}$ $6d^0$
99	Einsteinium	[Rn] $7s^2$ $5f^{11}$ $6d^0$
100	Fermium	[Rn] $7s^2$ $5f^{12}$ $6d^0$
101	Mendelevium	[Rn] $7s^2$ $5f^{13}$ $6d^0$
102	Nobelium	[Rn] $7s^2$ $5f^{14}$ $6d^0$
103	Lawrencium	[Rn] $7s^2$ $5f^{14}$ $6d^1$
104	Rutherfordium	[Rn] $7s^2$ $5f^{14}$ $6d^2$
105	Dubnium	[Rn] $7s^2$ $5f^{14}$ $6d^3$
106	Seaborgium	[Rn] $7s^2$ $5f^{14}$ $6d^4$

107 Bohrium	[Rn] $7s^2$ $5f^{14}$ $6d^5$
108 Hassium	[Rn] $7s^2$ $5f^{14}$ $6d^6$
109 Meitnerium	[Rn] $7s^2$ $5f^{14}$ $6d^7$
110 Darmstadtium	[Rn] $7s^2$ $5f^{14}$ $6d^8$
111 Roentgenium	[Rn] $7s^1$ $5f^{14}$ $6d^9$
112 Ununbium	[Rn] $7s^2$ $5f^{14}$ $6d^{10}$
113 Ununtrium	[Rn] $7s^2$ $5f^{14}$ $6d^{10}$ $7p^1$
114 Ununquadium	[Rn] $7s^2$ $5f^{14}$ $6d^{10}$ $7p^2$
115 Ununpendium	[Rn] $7s^2$ $5f^{14}$ $6d^{10}$ $7p^3$
116 Ununhexium	[Rn] $7s^2$ $5f^{14}$ $6d^{10}$ $7p^4$
117 Ununseptium	[Rn] $7s^2$ $5f^{14}$ $6d^{10}$ $7p^5$
118 Unonoctium	[Rn] $7s^2$ $5f^{14}$ $6d^{10}$ $7p^6$

Appendix D

Draft of My Original Preface

The first time I encountered the word "cyclotron" was around 1979, in my fifth grade from a school's science pamphlet. I still remember the word since then but I never really know what it was for. The next time I got acquainted with the cyclotron was in 2012, during the time I was writing this book. From the developments of the search for the Higgs boson, I became aware of the LHC (Large Hadron Collider) particle accelerator of CERN (Conseil Européen pour la Recherche Nucléaire/European Organization for Nuclear Research) located between Switzerland and France and the Tevatron particle accelerator of Fermilab located in Batavia, Illinois. So for me to write about particle physics is like waking up from a very long sleep and just learning what had transpired during the time I was not aware of this field of science. As you will read afterwards, I came from a really different direction to reach this field of study. This was how I came to write about it.

What drove me to solve the problems of physics? The same as Einstein did earlier in his life—to earn a living. Only in my case, physics had never been my line of study or work. I wrote a book because I have ideas on how to solve some of the problems of physics and to earn a living as a writer. I wrote a book because I believed that I have something to offer to science and the world.

I have been writing full time since the middle of 2006. After I finished writing my first book near the middle of 2009, I queried several literary agents. I also emailed several authorities on the subject of my book to have them read my manuscript with the idea of getting an endorsement from them that I could use to get an agent. To my consternation, I got no response from those authorities and all the literary agents I queried declined. I thought that the reasons nobody wanted to help me or represent my work was because my claims were just too unbelievable or I am an unknown and have no credential in the field that I am writing. As I have read, it's a typical response from the agents. It's a Catch-22: You have to be at least almost famous published to get noticed but you can't get noticed since you still have yet to be published. Worrying and in panic about my future, I decided to write on a more mainstream subject—in physics.

The first physics book that I wrote was about light. It is a subject that seems to be too simple for the physicists—a subject that is not so very sexy or manly at all compared to the physicists' mighty arsenal. My theory on light talks about rainbows and spectrum, which one might be inclined to think I might as well talk about fairies, elves, leprechauns, and unicorns—a far-removed subject for the physicists who are used to formulating big theories and using big machines to smash particles together. Ironically, the knowledge of light is one of the very foundations of physics. Without it, physics is in the dark.

When I wrote my first book on physics, I was trying to understand the formula $E=mc^2$, of how the light c was "trapped" in the atom or matter. Since then, from the increasing knowledge I had gained, I had learned that physics is lost and is groping in the dark. In physics, it is like the blind leading the blind. The empirical science is naked! Physics is living in fairytale. The physics that I once thought so formidable a subject is not so strong after all. I learned that even brilliant physicists could still stumble and fall. I had learned that the much-vaunted mathematics is an accurate tool only when the one using it uses it with common sense. That is for the user to discern if mathematics is leading him or her into the path of whimsical pseudoscience. For example, I had seen particles of light turned into a wave—like that of the water waves. Physicists truly believe that light has a dual wave-particle property, and this idea had wormed its way into the very heart of Quantum Mechanics. (In my unpublished book, on my theory on light, I had overthrown the Wave Particle Theory of Light and Wave-Particle Duality. I had even explained Einstein's Photoelectric Theory in a different and much simpler way.) Mathematics had brought the physicists into the Never Never Land: String Theory's extra dimensions, multi-universes (multiverse), and parallel universes. It seems that physicists have a hangover from the mind-bending gymnastics of Einstein's Theory of Relativity. The physicists are trying to imitate Einstein by formulating their theories to have the same complexities with his theories. Physicists thought that the simple nature had resorted to sleight-of-hand to achieve these complicated things that we observed.

In this backdrop, I had finished my first book on physics near the middle of 2011 after starting it near the end of 2009. I had this grandiose and idealistic idea that my work will be welcomed and helped as it will help physics and the physicists in their work. I emailed some physicists asking them to read my work with the intention of getting an endorsement from them so I could use it to submit to a local academic publisher or use it in my queries to agents. I experienced the same response as my first book—no response. Later, when I read the book *Physics on the Fringe* by Margaret Wertheim, I learned that that is how scientists actually respond—utter silence. (Well, that was a comforting thing for me to learn.) From what I had read, I had learned that you better have a good reason to wake up a physicist.

CERN's LHC, the world's largest and highest energy particle accelerator, started its construction in 1998. It started operating on September 10, 2008 with the successful circulation of the proton beam on its main ring. It was stopped nine days later due to some electrical problem, fixed, and was started again on November 20, 2009. CERN embarked on finding the Higgs boson, the particle that purportedly gives mass to other particles. (What gives the Higgs boson its own mass, another Higgs boson? At this time, I already had a rough idea of where the mass of the particles came from.)

I had also accidentally stumbled on a documentary movie, "The Atom Smashers," which showed the development in the Tevatron's search for the Higgs boson. It aroused my interest in particle physics. Since then, I had been following the developments in both the LHC and the Tevatron.

Refuting the existence of the Higgs particle was not the reason why I wrote this book. When I wrote my first book on physics, on my theory on light, I was a spectator to what was going on in the world of physics particularly in the developments of the researches of the particle accelerators. From that time until now, I had cringed at the scientific drama and dreaded the surreal incredulity of what was going on: the claims, the announcements, and the written articles on the search for the Higgs boson. The purported discovery of the Higgs boson (dubbed as the "God particle") with the 4.9 sigma certainty of its discovery was just too much for me to handle. *Yes, it is a particle within the range that their theory predicted. But no, it is not the particle that they had been searching. You just can't find a non-existent particle. The "God particle" does not exist.* The problem is not with the Standard Model but rather with the theory or theories behind it. I would like to shout to the physicists and the physics community, "The emperor has no clothes!" Instead, I had to contain myself and put myself to work to finish this book. Nobody listens to a anybody, especially if he is not even a physicist or has the venerable title of the "Ph.D." after a name, a "Dr." before a name. Early on, I had learned that the way to an argument is the solid written work where one's ideas are stated clearly so that anybody can argue on it in writing. That is, it is in accordance with the saying: Put your money where your mouth is. (Scientists will shred any idea they disagree with and they can be very straight-forward and blunt. It is probably why most of them respond in utter silence as a substitute for polite dismissal.)

Fermilab's Tevatron is having difficulty with its funding and several times it was hopeful it could get some reprieve. The perceived energy-challenged Tevatron finally had to stop its operation on September 30, 2011. Reasons given for it were due to budget cuts and because of the completion of the LHC in which the US is one of the contributors to its expense.

The path that I took in physics also led me eventually to the structure of an atom. That— I believe is more spectacular than the search for the source of mass of the particles. When I started reading physics books, I also learned of the search for the Theory of Everything. It didn't take me time to home in to that seemingly unsolvable endeavor. I am confident that if I set my mind into a problem, I will by all means solve it. And so, even as I was writing my first book on physics, I had already an idea of a three part book series: theory of light, theory of gravity, and theory of everything. They are supposed to be published one after the other as the first one contributes to the next one until the previous two are included in the third final book.

Actually, I have only some vague idea on how to go along writing about them. I have no idea if I can actually solve the problems on gravity or push through with my third book. I am not a physicist and before I even wrote my first book in physics, I had never thought I could even much less write a book. From the experience I got from writing my first book, what I only know, as always, is that when I put my effort into it, I was more surprised with what I had achieved. It also helped to settle a little doubt in the back of my mind on my ability to write a physics book in that I have no reputation to protect and so I have nothing to lose. What effect could a person do who had just accidentally wandered into the formidable world of physics?

And so I have decided to write another book in physics. I have to decide which one to write: the theory of gravity or the theory of everything. I have a simple decision to make: solve one of these problems in physics within a reasonable amount of time, which means to me, within one to two years—not within my lifetime, as the pressures in many areas of my life is mounting. (I have no job and I practically depended on my father for support.) I had roughly outlined both books, selecting the topic that would not require me a lot of images and drawings. So, I decided on the theory of everything, which is related to researches of the particle physicists. I can sense its urgency with the drumming of the progress in the LHC. The big machine was built and the public wants results and answers.

As is the case with all my books, I started with an overdose of confidence in my belief that I could write it fast and solve every problem that comes along the way. I had since then been experienced enough not to provide a specific date to say that I could finish my work "in such and such time" since every time I thought I could finish it "in such and such time," it usually does not happen. But as always, I used that pressure to drive me to accomplish what I had set out to do: To save the world of physics? No, just to finish my book and get it published. In reality, no sane physicists would say that they can solve the things that I had written, I am writing, or I am about to write— within their lifetime. The problem of finding the Higgs boson had been going on for more than 40 years now and I had just started my work formally in January 2012. (It just occurred to me to think how much time does writing a book figures in one's lifetime. The years add up. I have come a long, long way for a man who doubts he can write a book at all.) It's not even arrogance for people to claim they can solve these problems in physics; they are just being plain hopeful.

From my reading of the articles that were mostly from the Internet, I can sense the pressure for the LHC to produce results judging always from the vague references to the cost of its construction (about $10 billion). I mean, that should be enough to put immense pressure on CERN to produce results.

I had read in the Internet articles in the early part of 2012 that the LHC will finally settle the issue of the existence or non-existence of the Higgs boson and announce their findings near the end of the year. I have this idea then of heading them off, but to my surprise they timed their announcement much earlier on July 4, 2012. By then, I was still in the depth of things making sense and fleshing out my book. I have read, heard, and interpreted a somewhat inconclusive announcement, but the Internet articles and print publications seem to suggest otherwise. As it happened, Tevatron had accumulated data that seem to back the "putative" results. (That's "results" with an "s" was because the LHC got two bumps from its detector making some buzz that there are actually two Higgs particles.) The evidence was near a "five gamma" in the physicists' lingo—a sure thing. The LHC had discovered a new particle within the range they predicted (125–126 GeV) where the Higgs boson should show up—which a certain particle did. Is it actually the Higgs boson that showed up? (Did they finally open the champagne? Is the party in LHC? Is it over for Tevatron? There was even the talk of the Nobel Prize. I had known otherwise.)

Looking back, I really should be thankful for not getting any help in the publication of my first book as it drove me to write my next book. As it is, I am already under immense pressure in my personal life. The stress of writing worsens my diabetes but I also have to write and publish a book so that I could take care of my health—again a Catch 22. (I was wondering if my work will not pan out and I will give up and physics go its merry way lost in its own devices. I believed there is really no "Second Coming" for the long-awaited "New Einstein." That is, there is no "New Einstein.") I don't feel any resentment at all for those I had asked for help. After all, I was only asking for a great upheaval in physics—no less than a revolution that will also affect their work.

CERN announced on December 19, 2012 that there will be a definitive announcement coming March 2013 if they had indeed discovered the Higgs boson. By then, the LHC will go into a "2-year hibernation period" and will reopen in 2015 in the designed much higher energy to "unlock" more mysteries of the universe.

Meanwhile, I had been hard at work. I cannot hurry up my work just to chase and head off CERN. I will know when my work is finished—and by "finished," meaning that I am comfortable with my ideas and theories I had included in the book. I had learned to curb perfectionism in lieu of practicality and necessity. Anyway, the ball had been rolling since July 4, 2012; it is not mine to stop it. I had been burned twice before from asking help. Having no credential is not easy. They do have a leeway as they said their theory <u>predicted</u> the particle at a certain mass. That is, they can always seek comfort in that their "theory" predicted the mass of the particle.

December 21, 2012 came and went. The world did not end. I can hear a cosmic chuckle somewhere. But that's the least of my concern—I am looking forward to any result on the search for the Higgs boson.

Appendix D

On March 14, 2013, CERN had announced that what they had found was indeed the Higgs boson. The current developments in the Higgs boson is the argument for the renaming of the Higgs boson, which got its name from Peter Higgs, and that it should include its other contributors to the theory: Robert Brout, François Englert, Gerald Guralnik, C.R. Hagen, and Tom Kibble. There was also an argument on who should get the (possible) Nobel Prize as the Nobel Prize for a certain field is awarded only to a maximum of three individuals. There is the question as to who would get it or should the prize be shared with the team of scientists and engineers at the LHC.

The purpose of LHC and the search for the Higgs boson had taken its course. I don't know the excitement it had generated in the academic institutions and in the particle physics research facilities. Understandably, I don't feel any elation at all when they announced that they had found the Higgs boson. Rather, I felt some sense of sadness. If CERN had announced that the Higgs boson do not exist, my work will just support it. Since they had announced that that the Higgs boson exist, then the inevitability that my ideas will be pitted against the big machine has come to the fore and I have no doubt my theories will prevail. Let us see if the saying that "Nothing is more powerful than an idea whose time has come" is really true.

Physics is treading on the wrong path. The only way to steer the great ship of physics towards the right direction is to present new theories that are so robust that it can withstand common sense questions and practical observations. The new theories should be able to be tested by experiments, make predictions, explain wide variety of phenomena, and be in agreement with each other—in short, theories that are grounded in reality. It used to be that the physicists' approach to explaining the whole is to provide theories in parts. The problem is, those theories don't agree with each other at all. It is like a game of finding which theory or theories do not belong with the others.

In order for the new age of Real New Physics to rise, great theories must fall. In this day and age, there are a lot of formidable brilliant minds out there waiting for a great idea or theory they can espouse. I believed that physicists are not beholden to any ideas or theories that cannot pass muster the scientific and practical scrutiny. I believed that the physicists move forward with the strength of the scientific ideas or theories. Long before we know it, we are in the clasp of the Real New Physics—the Renewal Physics. (In the new age of physics, there is no "God particle" and no Big Bang.)

Notes and References

Chapter 1

1. Democritus <http://en.wikipedia.org/wiki/Democritus> (12 January 2012).

2. Georg Bauer is also known by his Latinized name, Georgius Agricola, being that Bauer means "farmer."

3. The man who invented the electron <http://ingeniousirelend/2011/02/the-man-who-'invented'-the-electron/> (27 April 2012).

4. George Johnstone Stoney <http://en.wikipedia.org/wiki/George_Johnstone_Stoney> (27 April 2012).

5. Corpuscles to Electrons <http://www.aip.org/history/electron/jjelect.htm> (27 April 2012).

6. Ibid.

7. The man who 'invented' the electron <http://ingeniousireland.ie/2011/02/the-man-who-'invented'-the-electron> (27 April 2012).

8. Hantaro Nagaoka <http://en.wikipedia.org/wiki/Hantaro_Nagaoka> (14 October 2009).

9. Benson, 829-830.

10. Hantaro Nagaoka <http://en.wikipedia.org/wiki/Hantaro_Nagaoka> (14 October 2009).

11. The Brownian motion was named after the Robert Brown (1773-1858), a Scottish botanist, who in 1828 had published his detailed observations of minuscule particles of pollen suspended in water can be seen to wiggle and wander when examined under the microscope.

12. The year 1905 was Einstein's Annus Mirabilis ("extra-ordinary year"). Einstein published three scientific papers in the same volume of the *Annalen Physik*, a German scientific journal: the photoelectric effect, Brownian motion and the atom, and the special theory of relativity.

13. Paul Ulrich Villard <http://en.wikipedia.org/wiki/Paul_Ulrich_Villard> (1 May 2012).

14. Benson, 830.

15. Fraser, 38.

16. Who Discovered Neutron and When <http://wiki.answers.com/Q/Who_discovered_neutron_and_when> (7 April 2012).

17. James Chadwick <http://en.wikipedia.org/wiki/James_Chadwick> (7 April 2012).

18. *New Book of Knowledge*, Volume A, 485.

19. Noel Giffin, "Size of the Atom," 23 January 1996, <http://trshare.triumf.ca/~safety/EHS/rpt/rpt_1/node7.htm> (5 April 2012).

20. How big is the nucleus in comparison to the entire atom? <http://wiki.answers.com/Q/How_big_is_the_nucleus_in_comparison_to_the_entire_atom> (5 April 2012).

21. How many atoms can fit inside a period? <http://wiki.answers.com/Q/How_many_atoms_can_fit_inside_a_period> (2 May 2012).

22. Proton <http://en.wikipedia.org/wiki/Proton> (23 April 2012).

23. Spin (physics) <http://en.wikipedia.org/wiki/Spin_(physics)> (5 September 2012).

24. What was Erwin Schrodinger's atomic theory? <http://wiki/answers.com/Q/What_was_erwin_schrodinger's_atomic_theory> (25 April 2012).

25. Nuclear shell model <http://en.wikipedia.org/wiki/Nuclear_shell_model>(22 April 2012).

26. Maria Goeppert-Mayer, the Nuclear Shell Model, and Magic Numbers <http://www.osti.gov/accomplishments/mayer.html> (12 September 2012).

Chapter 2

1. Quark <http://en.wikipedia.org/wiki/Quark> (19 October 2009).

2. Lederman, 315. Lederman's book, *The God Particle*, was a candid look at the field of particle physics. It was in this book that I got a glimpse of the glory and shortcomings of physics. While physics is led by many brilliant minds, physicists had succumbed to the academic discipline of following the known well-beaten path.

Chapter 3

1. In 1600, William Gilbert, who was the physician to Queen Elizabeth I coined the term *electric* from the Greek word *elektro*n.

2. Benson, 434.

Bibliography

Textbooks

Benson, Harris. *University Physics*. Canada: John Wiley & Sons, Inc., 1991.

Corwin, Charles H. *Introductory Chemistry: Concepts and Connections*. 2nd ed. New Jersey: Prentice Hall, 1998.

Giancoli, Douglas C. *Physics for Scientists & Engineers with Modern Physics*. 3rd ed. Upper Saddle River, New Jersey: Prentice Hall, 2000.

Knight, Randal D. *Physics For Scientists and Engineers with Modern Physics: A Strategic Approach.* Extended edition. San Francisco: Pearson Education, Inc., 2004.

Sears, Francis W., Mark W. Zemansky, and Hugh D. Young. *College Physics.* 6th ed. Canada: Addison-Wesley Publishing Company, 1985.

Young, Hugh D. and Roger A. Freedman. *Sears and Zemansky's University Physics with Modern Physics.* 11th ed. San Francisco: Pearson Education, Inc., 2004.

Books

Close, Frank, Michael Maren, and Christine Sutton. *The Particle Odyssey: A Journey to the Heart of the Matter*. New York: Oxford University Press, 2002.

Fraser, Gordon, Egil Lillestøl, and Inge Sellevåg. *The Search for Infinity: Solving the Mysteries of the Universe*. New York: Facts On File, Inc., 1995.

Fristzsch, Harald. *Quarks: The Stuff of Matter*. New York: Basic Books, Inc., 1983.

Kane, Gordon. *The Particle Garden: Our Universe as Understood by Particle Physicists*. Cambridge, Massachugb\hsetts: Perseus Publishing, 1995.

Lederman, Leon with Dick Teresi. *The God Particle: If Universe is the Answer, What is the Question?* New York: Mariner Books, 2006.

Oerter, Robert. *The Theory of Almost Everything: The Standard Model, the Unsung Triumph of Modern Physics*. New York: Plume, 2006.

Rothman, Tony. *Instant Physics: From Aristotle to Einstein, and Beyond*. New York: Byron Preiss Visual Publication, Inc., 1995.

Acknowledgements

There were unbroken circumstances and helps that made it possible for me to write my books. And so, I promised to myself when I published my first book that I will thank those who had made it somehow possible for me to accomplish my work. Without each one of them, I cannot be here to present my work. If my birth was timed for my reason of me being here, then these series of circumstances had been a great help to "pad" the time when my writings and the developments in physics had to coincide. Anything short of that and I would have missed this momentous event.

Looking back, the moment I came here in the US from the Philippines had been crucial in my personal education as I can say that I practically "grew up" here in the US. Here in the US I had the educational (self-taught), material (particularly books), and financial opportunity to pursue my interests.

I would like to thank my uncle, Filomeno M. Aduana, for having petitioned me for a working visa here in the US.

I would like to thank my father, Efren M. Aduana, for providing a place for me to stay and for supporting me financially all these years.

I would like to thank Luz Te and Virgilia Jao, both are nurse. Luz Te was the one who recognized my need to get a check-up in the Cook County Hospital, where I was diagnosed with Type 2 diabetes.

I would like to thank my past doctors and my present doctor, Dr. Irene Martinez of the John H. Stroger Jr. Hospital of Cook County in Chicago.

I would like to thank my brother, Gleen and his wife Joy, who had been a great help in providing a place for me to stay after my father retired.

I would like to thank my father and mother for providing part of their pension to help me continue to write my books.

I believe that success is nurtured first by the parents, then by the teachers who cared to teach, and then by the people and environment around him or her.

I would like to thank my mother who had taken a keen education of her four boys. She bought an encyclopedia when I was in fourth grade, which had started me to learn about places, arts, inventions, and famous luminaries of science. She's the strongest woman I had known, raising alone four boys and dealing with life's challenges.

The teachers' teachings are the building blocks of our education. Even if we forgot some of their names through the years, we owe them our gratitude. In my education, there were two teachings that I had not forgotten during high school. I would like to thank Mr. Cielito Fedoc, our Practical Arts teacher in my second year in high school. He taught us on how to draw simple technical drawings, which gave me a leg up in my technical drawing subjects later in engineering. My training in technical drawings had been helping me in drawing my inventions and the figures in my books. I would like to thank Mrs. Virginia Laguna, our Physics teacher in my fourth year in high school for teaching us how to understand an equation. That one simple formula for speed as the distance over time ($v = d/t$) had stuck in my mind all these years and helped me recognized and understood Einstein's $E = mc^2$.

In these times of information age, I get my information from the Internet, television documentaries, book stores, and the libraries. I would like to thank the Arlington Heights

Acknowledgements

Memorial Library (AHML), Arlington Heights, Illinois for my use of the library and whose staffs are always there to help.

These are the confluence of peoples, events, and time that had brought me and my work to the fore. It was not an easy path but I had stayed the course. I had finished what I had set out to do, what was set out for me to do.

Author's Note

It took more than forty years for the physicists to "discover" the Higgs boson—and it will take probably who knows how many more years for the physicists to discover that they were wrong without this book to point out that they were wrong. It took me about nineteen months to finish this book.

I had earlier set a "deadline" to finish this book, which forced me to find a freelance editor. Thankfully, she had backed out upon making a partial sample edit of the Preface. It's the same Preface which had given me a hard time, forcing me to place it at the back of the book. Having no editor had forced me and given me time to polish this book to the best of my ability. Although I had wished for an editor, that was way past my *self-imposed* deadline now. This led me also to the decision that if I cannot have an editor with a scientific or technical background in physics that it is better that I venture its self-publication instead. I was banking more on the importance of the information that my book brings more than anything.

There are various immense pressures that are pushing me to finish my book from the personal to the timely subject of the book itself to the date with history. Some of the reasons I had listed are the following:

- To achieve financial independence.
- To afford to pay for my health care.

- To reach a strategic position on the issue of my immigration status.
- To strike at the beginning of school year to prepare the students and teachers for the impending great changes in physics.
- To head-off the possible nomination for the Nobel Prize of the discovery of the Higgs boson.

Unbeknownst to anybody but me, there were many times when the world had nearly lost me. The only way for the world to learn the significance of my works is for me to publish whatever I have right now—even one book—in no matter what condition it is.

About The Author

Efren Basa Aduana Jr. is a graduate of Bachelor of Science in Electrical Engineering in the Philippines. He is an inventor and a writer. He had patented two simple inventions and had been writing full time since the middle of 2006.

He never had the chance to work in any field of electrical engineering. He never studied physics after college and is not in any way currently affiliated with any physics society, institution, or school.

He likes to read books, invent, cook, play the guitar, sing, listen to music, watch movies, carpentry, gardening, write poetry, drawing, photography, and biking. He would like to go back to inventing after writing all his books.

EfrenBasaAduanaJr@gmail.com

www.ingramcontent.com/pod-product-compliance
Lightning Source LLC
Chambersburg PA
CBHW060016210326
41520CB00009B/910